Acinetobacter baumannii - The Rise of a Resistant Pathogen

Edited by Karyne Rangel and Salvatore Giovanni De-Simone

Published in London, United Kingdom

Acinetobacter baumannii - The Rise of a Resistant Pathogen
http://dx.doi.org/10.5772/intechopen.1001504
Edited by Karyne Rangel and Salvatore Giovanni De-Simone

Contributors
Catherine Nonyelum Stanley, Amaka Marian Awanye, Ukamaka Chinelo Ogbonnaya, Amir Emami, Neda Pirbonyeh, Fatemeh Javanmardi, Eunice Damilola Wilkie, Jude Oluwapelumi Alao, Tosin Akin Akinmolayan, Anup R. Warrier, Sneha Radha, Alexandre Bueno Merlini, Renan Henrique Merlini, Bernardo Noya de Abreu, Maria Daniela Di Dea Bergamasco, Thiago Pavoni Gomes Chagas, Thiago Pavoni Gomes Chagas, Karyne Rangel, Salvatore Giovanni De-Simone

First published in London, United Kingdom, 2024 by IntechOpen
IntechOpen is the global imprint of INTECHOPEN LIMITED, registered in England and Wales, registration number: 11086078, 5 Princes Gate Court, London, SW7 2QJ, United Kingdom

British Library Cataloguing-in-Publication Data
A catalogue record for this book is available from the British Library

Additional hard and PDF copies can be obtained from orders@intechopen.com

Acinetobacter baumannii - The Rise of a Resistant Pathogen
Edited by Karyne Rangel and Salvatore Giovanni De-Simone
p. cm.
Print ISBN 978-0-85466-371-2
Online ISBN 978-0-85466-370-5
eBook (PDF) ISBN 978-0-85466-372-9

For EU product safety concerns:
IN TECH d.o.o., Prolaz Marije Krucifikse Kozulić 3, 51000 Rijeka, Croatia,
info@intechopen.com or visit our website at intechopen.com.

Meet the editors

Dr. Karyne Rangel graduated with a degree in Food Engineering from Estácio de Sá University (UNESA) in 2001. She also obtained an MSc and DSc in Sciences (Health Surveillance) from the National Institute for Quality Control in Health/ Oswaldo Cruz Foundation (INCQS/FIOCRUZ), in 2004 and 2013, respectively. She obtained a post-doctorate at INCQS/ FIOCRUZ and the Center for Technological Development in Health (CDTS)/ FIOCRUZ. For more than seventeen years, she has been dedicated to the field of medical microbiology with an emphasis on bacteriology, in particular *Acinetobacter baumannii*, developing studies in healthcare-related infections, molecular biology, antimicrobial resistance, multidrug resistance, integrons, virulence mechanisms, antimicrobial peptides, multilocus sequence typing, detection of antimicrobial resistance genes, and Pulse Field Gel Electrophoresis (PFGE). She is also developing a study on the effect and proof of ozone's effectiveness in the ESKAPE group's bacteria.

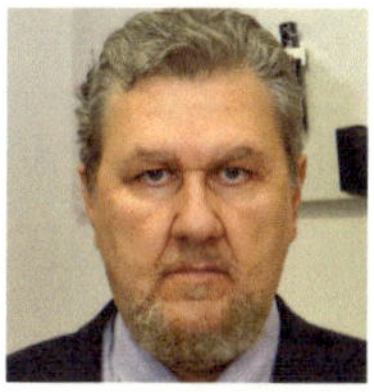

Salvatore Giovanni De-Simone obtained a BSc in Biomedical Sciences from the Rio de Janeiro State University, Brazil in 1976, and an MSc and DSc in Biochemistry from the Federal University of Rio de Janeiro, Brazil in 1980 and 1986, respectively. He was a visiting researcher at the Laboratory of Genetic Recombination, Institut d'Hematologie-Centre Hayem Hospital Saint Louis, France in 1987; Department of Medical Parasitology, New York University, USA in 1988–1989; and European Molecular Biology Laboratory, Germany in 1990. He was a Full Professor of Biochemistry at the Federal Fluminense University (UFF), Brazil in 1984–2012. He was Chief of the Biochemistry of Protein and Peptides Laboratory and Coordinator of the Peptide Synthesis Platform at Oswaldo Cruz Foundation (FIOCRUZ), Brazil, from 1992 to 2012. He is a permanent member of the Post-Graduation Program in Science and Biotechnology, UFF, and the Post-Graduation Program in Parasitic Disease, FIOCRUZ. He is the primary advisor of more than forty master's and thirty doctorate candidates and more than twenty-seven visiting researchers. Dr. De-Simone coordinates seventeen national and international symposiums and courses and thirty-five research projects sponsored by national and international agencies. He is an advisor of Brazilian universities and projects for several national and international agencies and an ad hoc reviewer for more than fifty international journals. He is also a guest editor of special issues of the International *Journal of Molecular Science* and *Toxins*. Furthermore, Dr. De-Simone is a specialist in science and technology production and innovation in health for the Center for Technological Development in Health (CDTS), FIOCRUZ; a member of the National Institute of Neglected Population Disease (IDN-Pd) of Brazil; a scientist at the Carlos Chagas Filho Foundation for Research Support of the State of Rio de Janeiro (FAPERJ); and a researcher for the National Council of Scientific and Technological Development (CNPq|).

Contents

Preface **IX**

Section 1
Insights into *Acinetobacter baumannii*: A Review of Microbiological Resistance Traits Host-Pathogen Interactions and Treatment Options in a Threatening Nosocomial Pathogen **1**

Chapter 1 **3**
Acinetobacter baumannii: Epidemiology, Clinical Manifestations and Associated Infections
by Catherine Nonyelum Stanley, Amaka Marian Awanye and Ukamaka Chinelo Ogbonnaya

Chapter 2 **27**
The Battle against Antibiotic Resistance: Novel Therapeutic Options for *Acinetobacter baumannii*
by Amir Emami, Neda Pirbonyeh and Fatemeh Javanmardi

Chapter 3 **57**
Host-Pathogen Interactions in *Acinetobacter baumannii* Infections: Mechanisms of Immune Evasion and Potential Therapeutic Targets
by Eunice Damilola Wilkie, Jude Oluwapelumi Alao and Tosin Akin Akinmolayan

Section 2
Urinary Tract Infections Caused by *Acinetobacter baumannii* **75**

Chapter 4 **77**
Treatment of *Acinetobacter baumannii*
by Anup R. Warrier and Sneha Radha

Chapter 5 **101**
Current Options for the Treatment of Urinary Tract Infections Caused by Multiresistant *Acinetobacter baumannii*
by Alexandre Bueno Merlini, Renan Henrique Merlini, Bernardo Noya de Abreu and Maria Daniela Di Dea Bergamasco

Section 3
Toxicity of Polymyxins in *Acinetobacter baumannii* **113**

Chapter 6 **115**
Understanding the Harmful Impact of Polymyxins on
Acinetobacter baumannii
by Karyne Rangel, Thiago Pavoni Gomes Chagas
and Salvatore Giovanni De-Simone

Section 4
Acinetobacter baumannii in Latin America **143**

Chapter 7 **145**
Carbapenem-Resistant *Acinetobacter baumannii* in Latin America
by Thiago Pavoni Gomes Chagas, Karyne Rangel and
Salvatore Giovanni De-Simone

Preface

Acinetobacter baumannii, a superbug resilient against numerous antimicrobial agents, including last-resort options, has garnered immediate attention from public health authorities. The US Centers for Disease Control and Prevention (CDC) identifies *A. baumannii* as an imminent threat. At the same time, the World Health Organization (WHO) underscores the urgent need for research and development of antibiotics to address infections caused by this resilient pathogen. It has earned its place as a leading nosocomial "ESKAPE" pathogen, prompting the WHO to prioritize it as the foremost pathogen requiring urgently and critically needed new antibiotics.

Chapter 1, "*Acinetobacter baumannii*: Epidemiology, Clinical Manifestations, and Associated Infections", provides a comprehensive overview of the *Acinetobacter* genus, tracing its history to the emergence of *A. baumannii*. The chapter explores the global epidemiology of *Acinetobacter* infections, detailing outbreaks in healthcare facilities, communities, and diverse climates, including temperate and tropical regions, as well as conflict and disaster settings. It emphasizes the impact of climatic conditions on the prevalence of *A. baumannii*, its association with various diseases, and its role as a significant contributor to healthcare-associated infections worldwide. The chapter also addresses community-acquired infections arising from conflicts and natural disasters. It delves into clinical manifestations associated with *A. baumannii*, encompassing pneumonia, bloodstream infections, trauma, wounds, surgical site infections, endocarditis, meningitis, urinary tract infections, and other related conditions. It further explores virulence factors in *A. baumannii*, their role in pathogenesis, primary mechanisms of antibiotic resistance, and available treatment strategies.

Chapter 2, "The Battle Against Antibiotic Resistance: Novel Therapeutic Options for *Acinetobacter baumannii*", explores virulence factors contributing to *A. baumannii*'s pathogenesis and high mortality rates. The chapter delves into self-survival mechanisms, including outer membrane proteins, lipopolysaccharides, capsular polysaccharides, phospholipase, nutrient-acquisition systems, efflux pumps, protein secretion systems, quorum sensing, and biofilm production. It subsequently discusses antimicrobial drug resistance and strategies to overcome these challenges. Recognizing the limited treatment options and the failure of most antibiotics due to the spread of multidrug-resistant (MDR) bacteria, the chapter advocates for alternative treatment approaches, such as combined treatments, the reuse of existing medications, and explores potential therapeutic avenues like new antibiotics, bacteriophages, antimicrobial peptides, monoclonal antibodies, nanoparticles, gene editing, and others.

Chapter 3, "Host-Pathogen Interactions in *Acinetobacter baumannii* Infections: Mechanisms of Immune Evasion and Potential Therapeutic Targets", delves into the intricate interplay between *A. baumannii* and the host immune system. The chapter emphasizes the mechanisms employed by *A. baumannii* to escape and subvert the immune response, leading to persistent and challenging infections. It highlights key aspects of the host immune system, including innate and adaptive immunity, pattern recognition receptors, and immune cell responses, within the context of *A. baumannii* infections. The chapter also discusses virulence factors and strategies utilized by *A. baumannii*, such as biofilm formation and quorum sensing. It identifies potential therapeutic targets, including novel antimicrobial agents, immunotherapies, and host-targeted therapies.

Chapter 4, "Treatment of *Acinetobacter baumannii*", provides an overview of traditional treatment options and drug selection in MDR infections, supplemented by a brief review of the evidence. It also explores emerging treatment options.

Chapter 5, "Current Options for the Treatment of Urinary Tract Infections Caused by Multidrug-Resistant *Acinetobacter baumannii*", addresses the specific challenges posed by urinary tract infections caused by MDR *A. baumannii*, considering the limitations imposed by urinary medication concentrations. The chapter covers epidemiology, main risk factors, clinical manifestations, diagnosis, and treatment.

Chapter 6, "Understanding the Harmful Impact of Polymyxins on *Acinetobacter baumannii*", scrutinizes the resurgence of interest in polymyxins as a last-resort treatment for *A. baumannii*. It emphasizes the limited clinical options of polymyxin B and colistin and details challenges associated with their administration, including high toxicity, notably nephrotoxicity, and neurotoxicity, along with less common adverse effects like injection pain, hypersensitivity reactions, and bronchospasms.

Chapter 7, "Carbapenem-Resistant *Acinetobacter baumannii* in Latin America", sheds light on the escalating prevalence of *Acinetobacter* infections globally, focusing on Latin America. The chapter details the widespread occurrence of carbapenem-resistant *A. baumannii* (CRAb), which poses a significant threat to public health. It outlines the high carbapenem resistance rates in *A. baumannii* worldwide, with Latin America experiencing some of the highest rates globally. The review summarizes the distribution of CRAb and its primary resistance mechanisms to carbapenems in Latin America.

Overall, this comprehensive assessment of *A. baumannii* offers a detailed exploration of its epidemiology, clinical manifestations, immune interactions, treatment options, and regional considerations, providing a valuable resource for researchers, clinicians, and policymakers grappling with the challenges posed by this resilient and problematic pathogen.

Karyne Rangel
FIOCRUZ, Center for Technological Development in Health (CDTS),
National Institute of Science and Technology for Innovation in Neglected Population Diseases (INCT-IDPN),
Rio de Janeiro, RJ, Brazil

Salvatore Giovanni De-Simone
FIOCRUZ, Center for Technological Development in Health (CDTS),
National Institute of Science and Technology for Innovation in Neglected Population Diseases (INCT-IDPN),
Rio de Janeiro, RJ, Brazil

Department of Molecular and Cellular Biology,
Biology Institute, Universidade Federal Fluminense (UFF),
Niterói, RJ, Brazil

Section 1

Insights into *Acinetobacter baumannii*: A Review of Microbiological Resistance Traits Host-Pathogen Interactions and Treatment Options in a Threatening Nosocomial Pathogen

Chapter 1

Acinetobacter baumannii: Epidemiology, Clinical Manifestations and Associated Infections

Catherine Nonyelum Stanley, Amaka Marian Awanye and Ukamaka Chinelo Ogbonnaya

Abstract

Acinetobacter baumannii is a Gram-negative, non-flagellated bacterium belonging to the coccobacillus family that is readily found in the environment. It has rapidly evolved, from an apparently innocuous organism to an opportunistic pathogen causing infections in both the hospital and the community. *A. baumannii* has attained the status of a superbug being resistant to many, including the last-resort antimicrobial agents, such as carbapenems, colistin and tigecycline. The Centers for Disease Control and Prevention (CDC) has classified *A. baumannii* as an immediate threat to public health, while the World Health Organization (WHO) is calling for research and development of critically needed antibiotics to treat these infections. It has earned a place as one of the most problematic nosocomial 'ESKAPE' pathogens causing the WHO to designate it as first on the list of pathogens for which new antibiotics are urgently and critically needed. *A. baumannii* has several mechanisms with which it is able to develop resistance to different antibiotics. It persists in hospital environments due to its ability to form biofilms and resist drying and disinfection. There is genetic diversity among the isolates of *A. baumannii*, thus making the study of this organism even more complex and underscoring the importance of sustained surveillance and good antibiotic stewardship to safeguard the publics' health.

Keywords: *Acinetobacter baumannii*, epidemiology, clinical manifestations, resistance, carbapenems, nosocomial infections

1. Introduction

It was the Dutch microbiologist Martinus Willem Beijerinck who in 1911 began the story of the genus *Acinetobacter*. Using medium enriched with calcium acetate, he had isolated a new organism from soil and named it *Micrococcus calcoaceticus* [1]. He named the organism but without a proper description of it, his report was largely disregarded and the name he proposed poorly accepted [2]. Over time, the same organism was described by other researchers under different names, some of which are listed as follows: *Micrococcus calcoaceticus*, *Moraxella lwoffi*, *Achromobacter mucosus*, *Alcaligenes haemolysans*, *Diplococcus mucosus* and *Neisseria winogradskyi* [1].

Brisou and Prevot coined the name *Acinetobacter* in 1954 to denote non-motile microorganisms in an attempt to differentiate between motile and non-motile members in the genus *Achromobacter* [3]. Paul Baumann and co-workers in 1968 did a comprehensive survey and proposed that the species listed above were of a single genus with similar phenotypic properties and hence did not need to be further divided and then proposed the name *Acinetobacter* [2, 4]. The genus *Acinetobacter* became officially accepted in 1971 following the work done by the subcommittee on the naming of *Moraxella* and related bacteria [4]. In 1974, Bergey's Manual of Systematic Bacteriology listed the genus *Acinetobacter* and further described it as a single species known as *Acinetobacter calcoaceticus*.

The *Acinetobacter* are Gram-negative, non-motile and non-fermenting strict aerobes. They are catalase-positive, oxidase-negative and non-fastidious bacteria whose DNA Guanine + Cytosine content ranges between 39 and 47% [1]. The genus *Acinetobacter* belongs to the *Moraxellaceae* family, *Pseudomonadales* order and is of the Gammaproteobacteria class of bacteria. Currently, about 74 species of *Acinetobacter* have been validated [5, 6]. *A. baumannii* is a highly ubiquitous and opportunistic coccobacillus with an extensive environmental spread. It has reservoirs in almost every environmental niche [5]. Although *A. baumannii* can be found in diverse milieus, such as soil, water, crude oil, sewage, inanimate objects and surfaces, skin and soft tissues, meat and dairy products and vegetables, among others, it thrives mostly in hospital environments [7]. Notwithstanding that the whole genome of *A. baumannii* had been sequenced in 2007 by Smith and co-workers (strain ATCC 17978) [8], its routine laboratory identification remains challenging because of the phylogenetic relatedness of the bacterium to many other species of the genus *Acinetobacter* known collectively as the *A. baumannii-calcoaceticus* (ABC) complex [1, 5, 7].

The species that originally constituted the *Acinetobacter baumannii-calcoaceticus* complex (previously called genomic species) are namely: *Acinetobacter baumannii*, *Acinetobacter pittii* (previously called genomic species 3), *Acinetobacter nosocomialis* (genomic species 13TU), *Acinetobacter seifertii*, *Acinetobacter lactucae* (also called *A. dijkshoorniae*) and *Acinetobacter calcoaceticus.* These species of *Acinetobacter* all belonged to the ABC complex [9]. These species are very difficult to distinguish phenotypically and share a very close genetic relatedness that makes molecular methods necessary for their identification [10]. Besides *A. calcoaceticus* whose pathogenicity is still somewhat unclear, other members of the ABC complex are established human pathogens. *A. baumannii* is the pathogen most frequently implicated in healthcare-associated infections (HAIs), with *A. pittii* and *A. nosocomialis* following closely [11].

Although *A. baumannii* was initially reasonably susceptible and responded well to antibiotic monotherapy, the bacterium has steadily demonstrated increasing rate of antibiotic resistance over the years [12]. This problem of increasing multidrug resistance (MDR) led the Infectious Disease Society of America (IDSA) to designate a group of bacteria consisting of *Enterococcus faecium*, *Staphylococcus aureus*, *Klebsiella pneumoniae*, *A. baumannii*, *Pseudomonas aeruginosa and Enterobacter spp.* as 'the ESKAPE pathogens' as a result of their ability to evade killing by antibiotics [13]. *A. baumannii* has become a superbug having developed resistance to virtually all known antibiotics in clinical use such as the fluoroquinolones, aminoglycosides and even the last-resort broad-spectrum carbapenems [13]. Multidrug-resistant *A. baumannii* (MDRAB) has attained a global epidemiology and is being encountered in hospital environments across the globe. Carbapenem-resistant *A. baumannii* (CRAB) had posed such a great burden on the healthcare system [13] that the World Health Organization (WHO) in 2017 listed it as a critical priority bacterium requiring the urgent development of new antibiotics [11]. Clinical outbreaks of *A. baumannii* infection have occurred in virtually every region of the world, with rates ranging between 1 and 30% with a greater burden on Eastern Europe [11].

The Middle East region has suffered a considerable number of outbreaks of MDRAB, for which it earned the "Iraqibacter" title [14]. These outbreaks of the bacterium were encountered during the Iraq war among US military hospitals in Iraq, Afghanistan and Kuwait [14]. In spite of extensive and continuing researches conducted to understand the antibiotic resistance mechanism of the bacterium, a clear and comprehensive understanding of the pathology, epidemiology and MDR mechanism of *A. baumannii* remains elusive. Added to these concerns are the very limited options available for the treatment of MDRAB infections. Until recently, Colistin was the only antibiotic still exhibiting therapeutic efficacy on strains of MDRAB, thus making its treatment very difficult with very limited therapeutic options [11]. Unfortunately, there have been reports on the emergence and continued rise of colistin-resistant *A. baumannii* in recent years [15–17], bringing to the fore the critical and urgent need for new antimicrobials, alternative treatment strategies, stricter infection prevention and control, and institution of rational antibiotic stewardship programmes. This chapter focuses majorly on the discussion of the current epidemiology of *A. baumannii*, its clinical manifestations and associated infections.

2. Species identification

Acinetobacter baumannii can be identified by cultural growth characteristics, biochemical characterization and molecular methods. They are classified as aerobic, Gram-negative, oxidase-negative, catalase-positive, indole-negative, urease-negative, haemolysis-negative, non-motile and non-lactose fermenting rods. Although it lacks flagella, *A. baumannii* can move along wet surfaces in an intermittent and jerky manner called twitching motility. It is non-fastidious and is easily grown in the laboratory on solid media such as sheep blood agar at an optimum temperature of 37°C. Growth can also occur at temperatures as high as 44°C. On blood agar, the colonies are about 1–2 mm in diameter and appear whitish, smooth or mucoid when the capsule is present. When grown on MacConkey agar, the colonies are light lavender in color, indicating non-lactose fermenting [18]. Some molecular methods that have been used for the identification of *A. baumannii* include restriction analysis and sequence analysis of the 16 s ribosomal RNA (rRNA) gene, ribotyping and transfer RNA (tRNA) spacer fingerprinting [19–23].

3. Epidemiology of *Acinetobacter baumannii*

Acinetobacter infections have a broad and global epidemiology. They have been implicated in outbreaks in both healthcare facilities and in the community; in temperate and tropical climates as well as in conflict and disaster situations [24–26]. Besides water and soil that constitute their natural milieu, they may also be found in pets, insects and other edible animals.

Common sites colonized in humans include the skin and soft tissues, blood, urinary, respiratory and digestive tract, wounds and the central nervous system [1].

The organism is also capable of surviving in biofilms from where it can migrate to the lower respiratory tract and trigger a pneumonia infection [27]. It is a common pathogen in the intensive care units (ICUs) and is associated with hospital-acquired pneumonia (HAP) and ventilator-acquired pneumonia (VAP) in patients with prolonged hospital stay. Tubing and other equipment involved in artificial ventilation can serve as a source of *A. baumannii* infection and result in lower respiratory tract infection. They are also responsible for other hospital-acquired infections (HAIs), such as wound infections, pressure ulcers, burn infections, septicaemia, urinary tract infections (UTIs), secondary meningitis and infective endocarditis [28]. *A. baumannii* is associated with skin and soft tissue infections and has been reported in traumatic injuries and postsurgical wounds.

Among members of the ABC complex, *Acinetobacter baumannii* has emerged as the best recognized and very important pathogen responsible for healthcare-associated infections as a result of its ability to survive under harsh environmental conditions. Its capacity to tolerate drying and thrive in the presence of minimal nutritional conditions confers on it a significant ability to acquire different mechanisms, which it uses to acquire resistance to various antimicrobial agents and to enhance its transmission in the healthcare setting [29, 30]. Complications arise leading to difficulty in treatment when burns get infected with *A. baumannii*. In some cases, a systemic infection can arise when the bacteria enter the bloodstream, leading to septicaemia. Prolonged use of catheters and antibiotic therapy have also been linked to *A. baumannii* infections. Although carbapenems, one of the broad-spectrum β-lactams with very high *in vitro* activity, used to be a preferred choice for treating infections due to *A. baumannii*, their clinical efficacy has suffered serious decline over time due to increasing resistance of the organism [31]. The organism had earlier developed resistance to many classes of antibiotics, such as β-lactam antibiotics, cephalosporins, aminoglycosides and fluoroquinolones. Only very few drugs, such as polymyxin B, colistin and tigecycline, are currently effective for MDRAB [32]. These drugs are expensive and are not readily available in resource-limited countries. Due to the high demand for colistin in the treatment of CRAB infections, colistin resistance has also been reported worldwide [33]. Resistance to polymyxin B and tigecycline has also been reported [34, 35].

Rates of carbapenem resistance differ according to geographic regions. The SENTRY Antimicrobial Surveillance Program observed that among *Acinetobacter* isolates collected between 2013 and 2016, susceptibility to meropenem was the lowest in Latin America (13.7%). This was followed by the Asia-Pacific region, Europe and the United States of America with 21.0, 22.2 and 54.9%, respectively [36]. In another study by Seifert and co-workers between 2016 and 2018, among *Acinetobacter* isolates collected, susceptibility to meropenem was the lowest for Africa and the Middle East (17.2%), closely followed by Latin America (19.6%). The susceptibility rates for Asia-South Pacific, Europe and North America were 31.4, 33.8 and 63.6%, respectively [37]. The WHO and the European Centre for Disease Prevention and Control (ECDC) in their latest report showed that year 2020 witnessed a wide variation in the percentages of carbapenem-resistant *Acinetobacter spp.* across Europe. Out of 38 countries and areas presenting data, less than 1% occurrence rate was seen in three countries, while the occurrence rate was 50% or greater in 35 others. Ireland, the Netherlands and Norway were the countries with the lowest rates, while in 21 countries, particularly in Southern and Eastern Europe, carbapenem resistance rates were as high as 50% or greater [31]. By means of molecular typing of *A. baumannii* isolates, the dissemination of three lineages of the organism in Europe was established and classified as European clones I, II and III [38]. These were later renamed international clones I, II and III (IC1, IC2 and IC3) in recognition of the fact that these lineages had already been disseminated globally [39, 40]. Nine international clones are presently recognized [41]. Two multilocus sequence type (MLST) schemes, known as Oxford and Pasteur, have also been used to characterize *A. baumannii* [31]. IC1 and IC2 are the most widely spread clones globally and often express the acquired carbapenemase oxacillin-hydrolysing (OXA)-23 [25, 26]. Regional variations do occur, with international clone V (IC5) and international clone VII (IC7) being more prevalent in Central and South America while international clone IX (IC9) is more prevalent in Africa and the Middle East [38, 41].

A higher CRAB colonization or infection was also observed in COVID-19 intensive care unit (ICU) in two studies in Italy. Another Italian study of 16 ICUs in the Piedmont area during the COVID-19 outbreak showed that 19% of COVID-19 infected patients became colonized or were infected by CRAB during their ICU stay, leading to a 67% mortality rate [42]. The United States of America, Argentina, Europe, Brazil, Japan, China, Hong Kong, Taiwan, Korea, Middle East, Nigeria and other African countries are areas where several epidemiological studies have documented the occurrence of infections due to *A. baumannii* [43, 44]. Some tropical regions of the world have experienced community-acquired pneumonia, particularly during warm and humid months [45]. An increase in the number of MDRAB was witnessed among the United Kingdom (UK) and US military personnel injured while on deployment to Iraq and Afghanistan [46].

The Middle East has also had its fair share of *A. baumannii* infection. MDRAB has been severally documented in hospitals in the United Arab Emirates (UAE), Bahrain, Saudi Arabia, Palestine and Lebanon and Egypt [47, 48]. In a retrospective study conducted to evaluate the prevalence of MDRAB responsible for infections in patients admitted at the ICU of the Riyadh Military Hospital, Saudi Arabia, *A. baumannii* was the most common bacterium isolated, representing 40.9% of the samples [39].

In Nigeria and other resource-limited countries in Africa, there is paucity of information regarding the molecular epidemiology and antimicrobial resistance status of *A. baumannii*, mainly due to lack of capacity for the isolation, identification and testing of antimicrobial susceptibility of these organisms. Nonetheless, a number of studies have been done to establish the molecular characteristics of *A. baumannii* isolates in Nigeria. In one such study conducted in Southwest Nigeria, the genetic diversity and molecular mechanisms of CRAB isolated from hospitals were characterized. All *A. baumannii* isolates submitted to the antimicrobial resistance surveillance reference laboratory in Nigeria between 2016 and 2020 had their genomes sequenced [49]. Eighty-six (86) *A. baumannii* isolates recorded belonged to 35 different Oxford sequence types (Oxf STs) and 28 Institute Pasteur STs (pas STs). Sixteen of the 35 distinct Oxford sequence types were novel. Thirty-eight of the isolates did not belong to any previously known international clone and more than half of the isolates expressed phenotypic resistance to 10 of the 12 tested antimicrobial agents. Fifty-four of the isolates were resistant to carbapenem, especially the IC7 and IC9 strains. In summary, the study recorded an increase in blaNDM-1 prevalence with widespread transposon-mediated dissemination of carbapenemase genes in different *A. baumannii* lineages in Nigeria's Southwest region. Other studies in the same region also found MDRAB with widespread carbapenemase resistance [49–51].

Some strains of the genus *Acinetobacter* have developed mechanisms that enable them to survive for long periods under harsh environmental conditions. This ability to withstand adverse conditions promotes their transmission in healthcare settings through contaminated fomites [52, 53].

3.1 Climatic conditions

Originally, *Acinetobacter* was more prevalent in tropical environment and was recorded as causing 17% of pneumonias associated with ventilator use in the ICU of a Guatemalan hospital. Only *Pseudomonas* had a higher prevalence than *Acinetobacter* in that study with 19% [54]. Over the last 5 decades, members of the genus *Acinetobacter* have evolved to become frequent nosocomial pathogens of concern even in the temperate regions [55]. This evolution of *Acinetobacter* from a little known apparently innocuous organism to an opportunistic pathogen credited with causing infections in both the hospital and the community has been linked to their possession of several survival mechanisms and their ability to rapidly develop resistance to most available antibiotics [56]. HAIs due to *Acinetobacter* have been established to be more prevalent in the tropical weather in summer compared to other seasons [57]. Between 1987 and 1996, the CDC received a report that reviewed 3447 cases of infections involving *Acinetobacter*. The rates of infection were established to be about 50% more from July to October compared to other seasons of the year. This increase was thought to be probably due to higher humidity of the air and contaminants suspended and transmitted in the air as aerosols. It is worthy of note that the condensate from air-conditioning units has been found to predispose to epidemic *Acinetobacter* infections [57].

3.2 Disease associations

Acinetobacter attained global prominence as a major cause of nosocomial infections. Patients in intensive care and those with compromised immunity were most vulnerable to *Acinetobacter* infection. *Acinetobacter* infections have however not been confined to healthcare settings alone. Cases of community-acquired infections due to *Acinetobacter* were reported in Australia and Asia. Outbreaks of *Acinetobacter* infections were also reported among soldiers during the war in Iraq [14].

3.3 Nosocomial infections

Acinetobacter has established itself as a prominent cause of healthcare-associated infections worldwide. The National Healthcare Safety Network (NHSN) in a 2016 report evaluated the prevalence of antimicrobial-resistant pathogens associated with nosocomial infections in the United States of America [58]. Among the frequent Gram-negative isolates, *Acinetobacter* species accounted for 12.8 and 8.8%, respectively, for VAP infection isolates and central line-associated bloodstream infection isolates, while catheter-associated urinary tract infection isolates and surgical site infection isolates accounted for 1.3% each.

Patients in the ICU, particularly the young and the elderly, and those in long-term care settings are more susceptible to *A. baumannii* [58]. Other factors that may predispose patients to infections with *A. baumannii* include recent surgery, catheter use, tracheostomy, artificial ventilation, parenteral nutrition and treatment with broad-spectrum antibiotics like carbapenem, fluoroquinolones and ceftriaxone [59, 60]. For neonates, low birth-weight, parenteral feeding and catheter use may pose added risks [61, 62]. Outbreak investigations are a primary data source for information about healthcare-associated *Acinetobacter* infections [54].

There have also been *Acinetobacter* outbreaks traceable to common-source contamination such as air conditioner or contaminated ventilator [57]. Cross-infection by healthcare workers caring for colonized or infected patients who do not maintain proper aseptic techniques including hand washing and touching inanimate objects can also lead to infection outbreak [43, 45, 54]. Introduction of *Acinetobacter* into a hospital may lead to serial or overlapping outbreaks due to MDR strains often seen at such times. Multiple strains, which may become endemic, are established with a single endemic strain being prevalent subsequently [54]. Protracted colonization may enhance endemicity of *A. baumannii* following an outbreak. In one study, colonization persisted for up to 42 months and affected 17% of patients [46]. Multicenter outbreaks have been recorded across the globe, in the United States of America, Europe, South America, Africa, Asia and the Middle East [47, 48, 60]. In 2005, there was an outbreak of carbapenemase producing (OXA-40) *Acinetobacter* in Greater Chicago area. Several hospitals and long-term facilities were affected along with many patients [63].

Several factors can lead to monoclonal outbreaks happening in multiple hospitals. These may be spread between institutions, via movements of patients or personnel, or exposure to common-source contamination of food or equipment. For this reason, the importance of regular epidemiological surveillance as well as infection prevention and control measures to stop the transmission and spread of *Acinetobacter* in long-term care facilities cannot be overemphasized. There is a paucity of data with respect to the prediction of patients suffering from infections due to *Acinetobacter*. Although mortality rates may be high among such patients, it cannot be said with certainty that the mortality was due to *Acinetobacter* infection [64]. For example, the consequence of *Acinetobacter* infections on mortality was indeterminate in a paired cohort study of patients with trauma [65]. Compared to control patients who had other infections that were not *Acinetobacter*, a longer stay in the ICU and increased organ failure were observed among cases exposed to *Acinetobacter*. Resistance to imipenem, compromised immunity as seen in old age and diabetes mellitus, the female gender and septic shock constitute some of the risk factors that cause mortality in those suffering from *Acinetobacter* infections [52, 53].

3.4 Community-acquired infection

There have been reports of community-acquired *Acinetobacter* infection in Australia and Asia [26]. In Australia, pneumonia occurring in the community was more prevalent during the rainy season [66]. In northern Australia with tropical climate, *A. baumannii* was implicated in 10% of cases of severe pneumonia acquired in the community [67]. Infections acquired in the community have been distinguished by pharyngeal presence of the organism, pneumonia that is aggressive and high case fatality rates. Chronic obstructive pulmonary disease, alcoholism, tobacco use, diabetes and cancer are some noted risk factors [66, 67]. Bloodstream infections have also been reported [66, 68].

Between February 2012 and October 2013, Rafei and co-workers in Lebanon conducted a study to evaluate the epidemiology of *A. baumannii* in the community outside the human body. Using cultural methods, they tested for the presence of *A. baumannii* in different samples covering the environment, water, food and edible animals. Species were identified using *rpoB* gene sequencing and antibiotic susceptibility was evaluated.

The *A. baumannii* isolates were studied using two genotyping approaches, namely multilocus sequence typing (MLST) and $bla_{\text{OXA-51}}$ sequence-based typing (SBT). Varying amounts of *A. baumannii* were isolated in all the samples. But for one isolate that expressed a $bla_{\text{OXA-143}}$ gene, all isolates were phenotypically susceptible to antibiotics tested and harbored no carbapenemase-encoding genes. Using MLST, 36 sequence types (STs) were obtained, with 24 of them being novel STs reported for the first time. The *bla*$_{OXA-51}$ SBT demonstrated the presence of 34 variants; 21 of them were novel and all were of animal origin. Human genotypes such as international clones I and X (IC1 and IC10) were detected in water and animals and the possible involvement of these new animal clones in human disease poses a public health concern. The researchers then concluded that animals could serve as the possible reservoir for *A. baumannii* and the spread of new emerging carbapenemases to humans [69]. The report of community-acquired infections is rare in the United States of America. Although the reason for the greater prevalence of *Acinetobacter* infections in certain regions has not been fully elucidated, it does appear to be connected to climatic differences that drive bacterial colonization.

3.5 Conflicts and natural disasters

Acinetobacter infections have been established as a common feature in wars and conflict situations. A rise in infections due to this organism has been reported several times in different conflicts such as in Korea, Vietnam, Iraq and Afghanistan, leading to the suggestion for it to be added as part of differential diagnosis of infections encountered among soldiers in combat and after natural disasters in a tropical region [14, 70]. In a study involving US troops stationed in Iraq and Afghanistan between 2007 and 2008, *A. baumannii* made up 63% of all bacteria isolated from soldiers' wounds [70]. In another report, *Acinetobacter* isolated from military personnel were less susceptible to imipenem than those isolated from individuals not actively engaged in war (63/87%) [70].

The genetic diversity and resurgence of *Acinetobacter* in personnel exposed to several military operations over many decades appear to suggest the involvement of multiple sources, such as local cuisines, contamination of wounds in combat, spread in the environment and cross-infection between the field and treatment centres [14, 70, 71]. *Acinetobacter* infections have also had an unusually high prevalence during natural disasters. During the 2004 tsunami in Southeast Asia, *Acinetobacter* resistant to several antibiotics were recovered from wounds, blood and respiratory fluids among 17 patients who sustained severe soft tissue injuries and fractures [72]. *A. baumannii* was also the most frequently isolated healthcare-associated pathogen in an ICU in Turkey after the 1999 Marmara earthquake in that country, despite having been only rarely isolated there previously [73].

4. Clinical manifestations of *A. baumannii* infections

Acinetobacter baumannii is of clinical importance partly due to its ability to survive in a broad range of temperatures and environmental conditions. It is highly resistant to desiccation and can survive for months on fomites. This makes them easy to spread in hospital settings where they can cause nosocomial outbreaks and contribute to the spread of MDRAB. Tubing and other equipment involved in artificial ventilation can serve as a source of *A. baumannii* infection and result in lower respiratory tract infection.

They are also responsible for other HAIs, such as wound infections, pressure ulcers, burn infections, septicaemia, UTIs, secondary meningitis and infective endocarditis [28]. *A. baumannii* is associated with skin and soft tissue infections and has been reported in traumatic injuries and postsurgical wounds. Complications arise leading to difficulty in treatment when burns get infected with *A. baumannii*. In some cases, a systemic infection can arise when the bacteria enter the bloodstream, leading to septicaemia. Prolonged use of catheters and antibiotic therapy have also been linked to *A. baumannii* infections.

Also, *A. baumannii* can develop resistance to many classes of antibiotics, such as β-lactam antibiotics, cephalosporins, aminoglycosides, fluoroquinolones and carbapenems. The increasing prevalence of CRAB and MDRAB has narrowed down therapeutic options making them a global concern. CRAB and MDRAB are associated with increased patient hospital stay and mortality [74]. Risk factors for high mortality include severity of infection, malignancy, older age, inappropriate use of antibiotics, renal failure, invasive procedures and prolonged stay in ICU [75, 76].

Only very few drugs, such as polymyxin B, colistin and tigecycline, are currently effective for MDRAB [32]. These drugs are expensive and are not readily available in resource-limited countries. Due to the high demand for colistin in the treatment of CRAB infections, colistin resistance has also been reported worldwide [33]. Resistance to polymyxin B and tigecycline has also been reported [34, 35].

5. *Acinetobacter baumannii*-associated infections

Gram-negative bacteria, such as *Acinetobacter baumannii, Pseudomonas aeruginosa, Klebsiella pneumoniae* and *Escherichia coli,* are common causes of many infections like pneumonia, bloodstream infections, wound and surgical site infections and meningitis in healthcare settings. *A. baumannii* is a common HAI and can infect various human anatomical sites. Clinical manifestations of *A. baumannii* infection are diverse, the most frequent ones being infections of the bloodstream and pneumonia associated with use of ventilators [77]. The severity and mortality rate of the infection and the patient outcome can depend on the virulence and antibiotic susceptibility of the infecting strain, such as MDRAB or CRAB, co-morbidities, length of hospital stay and other demographic characteristics [78].

5.1 Pneumonia

As *A. baumannii* can grow on a variety of environmental conditions, are tolerant to desiccants and can withstand many disinfectants and cleaning solutions, it is a common contaminant of hospital fomites. They are easily transferred from one patient to another or from a healthcare provider to a patient. Many of the hospital strains are resistant to many antibiotics due to constant exposure to antibiotics in the hospitals [79]. Hence, infections caused by these organisms are difficult to treat and are associated with longer periods of hospitalization. *Acinetobacter* infection is a major cause of pneumonia in patients in ICU who need assisted ventilation. A common feature of this pneumonia is delayed onset. In general, *Acinetobacter* pneumonia demonstrates other clinical manifestations that resemble those seen in pneumonia contracted in healthcare settings [24]. Pneumonia is a common and serious HAI, especially VAP, in patients in the ICU who are on artificial ventilation. Longer periods of antibiotic use, hospitalization and time on mechanical ventilators can increase the risk of *A. baumannii* infection [80]. Contaminated equipment and poor personal hygiene are common causes of transmission. In a prospective observational study conducted in nine countries in Europe across 27 ICUs, *A. baumannii* was established as one of the very common pathogens responsible for nosocomial pneumonia and it was actually the most prevalent isolate in Greece and Turkey [81]. Nosocomial pneumonia following *Acinetobacter* infection is linked with highly resistant isolates with mortality rates ranging between 35 and 70%.

5.2 Community-acquired pneumonia

Acinetobacter has been shown to cause severe community-acquired pneumonia that is distinguished by a stormy illness, in which the onset is abrupt and progression is rapid with resultant respiratory failure and uncertain haemodynamic parameters [24–26]. About a third of patients may experience septic shock. This situation, which appears to be more common in Australia and Southeast Asia when compared to other regions, has increasingly fatal outcome [82].

5.3 Bloodstream infections

Although bloodstream infections caused by *Acinetobacter* are responsible for a lower percentage of nosocomial infections, they are still a major public health concern since studies have revealed high mortality, especially in CRAB strains [83]. A retrospective observational study of bacteraemia caused by *Acinetobacter spp.* was undertaken in a UK hospital. *A. baumannii* was the most frequently isolated species. Most cases of bacteraemia occurred in patients in ICU and were associated with CRAB and MDRAB and these were associated with higher mortality rates, irrespective of appropriate empirical antibiotic therapy [84]. Vascular catheters and the respiratory tract are the most common sources of bacteraemia due to *Acinetobacter* infection [85, 86]. The urinary tract and wounds contribute to bloodstream infections to a lesser extent. Among factors that may predispose to *Acinetobacter* bloodstream infections are prolonged hospital or ICU stay, immunosuppression, trauma, burns, cancer, mechanical ventilation, previous surgery, previous use of broad-spectrum antibiotics, immunosuppression, trauma, burns, malignancy and invasive procedures [85–89].

5.4 Trauma, wound and surgical site infections

In healthcare settings, many organisms can infect the skin and soft tissue including *A. baumannii*. It has been associated with delay in wound healing, skin graft rejection and death from sepsis. Cases of skin and soft tissue infections caused by *A. baumannii* have also been recorded following a blast injury and chronic leg ulcer [90]. Fleming et al. in a wound infection mouse model demonstrated that iron depletion plays a crucial role in the pathogenesis of *A. baumannii* wound infections [91]. Exogenous supplementation of iron to the wound site prevented the activation of virulence genes involved in iron acquisition.

Contamination of surgical and traumatic injuries by *Acinetobacter* may result in severe infection of the soft tissue that may ultimately lead to osteomyelitis [92]. *Acinetobacter* is not commonly implicated in both community- and hospital-acquired skin infections like cellulitis and folliculitis [19, 93, 94]. MDRAB is, however, becoming more prevalent in injuries sustained during conflicts.

5.5 Endocarditis

Acinetobacter species have been implicated as a rare cause of infective endocarditis in people with artificial heart valves [89, 95, 96]. *Acinetobacter* was responsible for two cases of heart valve endocarditis due to nosocomial bacteraemia in a study that investigated 171 patients with prosthetic heart valve [97]. *Acinetobacter* endocarditis is typically characterized by acute onset with an aggressive course.

Mortality tends to be higher in the setting of native valve endocarditis than prosthetic valve endocarditis, likely because of the low index of suspicion leading to delayed treatment in such cases [96].

5.6 Meningitis

Nosocomial meningitis may sometimes result from *Acinetobacter* infection [98, 99]. Prior antibiotic therapy, neurosurgical procedures and intracranial hemorrhage are some risk factors for meningitis [100–102]. Outbreaks of nosocomial *Acinetobacter* meningitis were documented in the course of administering contaminated methotrexate via the intrathecal route [103]. Survivors of nosocomial meningitis may suffer severe sequalae [104]. Though *Acinetobacter* meningitis is not commonly encountered in the community, it does occur majorly in hitherto healthy individuals in the tropics and is often not resistant to drugs [105]. Common symptoms seen in *Acinetobacter* meningitis cases include fever and meningeal signs with seizures sometimes present. *Acinetobacter* central nervous system infections may present other clinical manifestations similar to those generally seen in meningitis.

5.7 Urinary tract infection

Acinetobacter can readily colonize the urinary tract, especially when there is an indwelling urinary catheter although the incidence of infection is low [83, 106]. A study in the United States of America reviewed 5000 urinary tract infections in medical ICU. Only 1.6% of the infections were attributed to *Acinetobacter* and 95% of these were linked to urinary catheters [83]. Urinary tract infection acquired in the community may occur very sparingly [107, 108]. In the absence of other signs or symptoms of infection, isolation of *Acinetobacter* may be attributed to colonization.

5.8 Other infections

Acinetobacter colonization has been reported in wearers of contact lens, and eye infections such as corneal ulcers may occur [109, 110]. In a study of 750 cases of corneal ulcers, *Acinetobacter* was the third leading cause, responsible for 7% of the cases [111]. The infections usually occurred after cataract or other eye surgeries.

Patients admitted to the ICU may develop nosocomial sinusitis due to *Acinetobacter* for which mechanical ventilation is a very important predisposing factor [112]· *Acinetobacter* sinusitis can progress to pneumonia since the infected sinuses serve as reservoirs for the organism, which can subsequently be disseminated to the lower respiratory tract [112].

6. Virulence properties

Acinetobacter baumannii is an opportunistic pathogen that has a high incidence among immunocompromised patients, especially those with a prolonged hospital stay. Virulence factors associated with the organism include an outer membrane protein A (OmpA), porin proteins, capsule formation, lipopolysaccharide (LPS) endotoxin, iron acquisition systems and biofilm formation [113]. OmpA is the most abundant surface protein on *A. baumannii* and contributes immensely to the pathogenic potential of the organism. It binds to receptors on the host cell surface,

Virulence factor	Role in disease
Porin proteins, e.g., OmpA	• Adherence and invasion
	• Induction of apoptosis
	• Serum resistance
	• Biofilm formation
Polysaccharide capsule	• Serum resistance
	• Survival in tissue infection
	• Evasion of host immune responses
	• Biofilm formation
Lipopolysaccharide (LPS)	• Serum resistance
	• Survival in tissue infection
	• Evasion of host immune responses
Outer membrane vesicle (OMV)	• Delivery of virulence factors
	• Horizontal transfer of antibiotic resistance gene
	• Evasion of host immune responses
Outer membrane proteins (OMPs), e.g., metal (Fe, Zn, Mn) acquisition systems; protein secretion systems (Types II and IV); penicillin-binding proteins, etc.	• *In vivo* survival
	• Killing of host cells
	• Host colonization
	• Biofilm formation
	• Serum resistance

Table 1.
Virulence factors in Acinetobacter baumannii *and their role in pathogenesis.*

thereby inducing apoptosis. It also mediates resistance to complement proteins and is involved in biofilm formation [114, 115]. These functions help the bacterium to grow under unfavorable conditions and survive both within and outside the host. Fimbriae, phospholipases C and D are other cell surface structures and proteins that contribute to the virulence property of *A. baumannii*. Fimbriae like OmpA are involved in adhesion to host cell surface and promote colonization. Phospholipase C is toxic to host epithelial cells, while phospholipase D mediates serum resistance, evasion of host epithelial cells and promotes disease pathogenesis [116]. The virulence factors identified for *A. baumannii* are presented in **Table 1**, adapted from [117].

7. Antibiotic resistance

Acinetobacter baumannii has intrinsic resistance to many antibiotics and also easily acquires resistant genes from other bacteria. Acquisition of antibiotic resistance is usually mediated via horizontal transfer of antibiotic genes from other organisms. Genome sequencing of some strains of *A. baumannii* revealed that resistant genes were acquired from species of *Pseudomonas, Escherichia* and *Salmonella* [24]. The major mechanisms of antibiotic resistance in *A. baumannii* are presented in **Table 2**.

Antibiotic class	Resistance mechanisms
β-Lactam	• β-Lactamase production
	• Carbapenemase production
	• Loss of outer membrane porin proteins
	• Efflux pump reduces antibiotic concentration inside the cell
	• Altered expression of penicillin-binding proteins (PBPs)
Tetracyclines	• Efflux pump reduces antibiotic concentration inside the cell
	• Ribosomal protection
Glycylcylines	• Efflux pump reduces antibiotic concentration inside the cell
Aminoglycosides	• Enzymatic degradation
	• 16 s rDNA methyltransferases
Quinolones	• DNA gyrase
	• Efflux pump reduces antibiotic concentration inside the cell
Chloramphenicol	• Efflux pump reduces antibiotic concentration inside the cell
Trimethoprim / Sulfamethoxazole	• Efflux pump reduces antibiotic concentration inside the cell
	• Dihydropteroate synthase inhibitor
	• Dihydropteroate reductase inhibitor
Macrolides	• Efflux pump reduces antibiotic concentration inside the cell
	• Polymyxins

Table 2.
Major mechanisms of antibiotic resistance in Acinetobacter baumannii.

8. Treatment strategies

Acinetobacter baumannii is intrinsically resistant to many antibiotics and is capable of acquiring resistant genes via horizontal gene transfer. This makes the treatment of infections caused by *A. baumannii* challenging to treat. Carbapenems are generally considered as the antibiotics of choice for treating *A. baumannii* infections due to their efficacy and favorable safety profile. Polymyxin B, colistin and tigecycline are other antibiotics that can be used in cases of CRAB. Unfortunately, resistance to polymyxin B, colistin and tigecycline has also been reported. MDRAB has necessitated the search for other options including new drug discovery. Pan-drug-resistant *A. baumannii* that is resistant to at least one agent in all classes of antibiotics has rarely been reported. The organism is usually sensitive to one or more antibiotics. Thus, efficient combination therapy with at least one agent from different classes of antibiotics is currently used in the treatment of *A. baumannii* infections [117].

9. Conclusion

In conclusion, *A. baumannii* has become established as a pathogen of global dimension that is prevalent in various environmental niches. As it has developed resistance to many antibiotics including those that were considered to be the last resort, treatment of infections caused by this organism has become a major challenge for clinicians. Efforts in the research and development of new antibiotics and treatment strategies are yet to yield novel results and hence the need to revisit traditional methods. Effective public health policies in both the community and hospital can help control *A. baumannii* infections. The saying that 'prevention is better than cure' still holds true, thus in addition to concerted efforts to develop new and alternative treatment strategies, stringent infection prevention and control mechanisms along with continuous epidemiological surveillance should be instituted to curtail the transmission and spread of MDRAB.

Author details

Catherine Nonyelum Stanley[1*], Amaka Marian Awanye[1]
and Ukamaka Chinelo Ogbonnaya[2]

1 Faculty of Pharmaceutical Sciences, Department of Pharmaceutical Microbiology and Biotechnology, University of Port Harcourt, Nigeria

2 Faculty of Allied Health Sciences, Department of Medical Laboratory Sciences, University of Port Harcourt, Nigeria

*Address all correspondence to: catherine.stanley@uniport.edu.ng

References

[1] Peleg AY, Seifert H, Paterson DL. *Acinetobacter baumannii*: Emergence of a successful pathogen. Clinical Microbiology Reviews. 2008;**21**(3):538-582. DOI: 10.1128/CMR.00058-07

[2] Henriksen SD. *Moraxella, Acinetobacter*, and the *Mimeae*. Bacteriological Reviews. 1973;**37**(4):522-561. DOI: 10.1128/br.37.4.522-561.1973

[3] Brisou J, Prevot AR. Studies on bacterial taxonomy. X. The revision of species under *Acromobacter* group. Annales de l'Institut Pasteur. 1954;**86**(6):722-728

[4] Baumann P, Doudoroff M, Stanier RY. A study of the *Moraxella* group. II. Oxidative-negative species (genus *Acinetobacter*). Journal of Bacteriology. 1968;**95**(5):1520-1541. DOI: 10.1128/jb.95.5.1520-1541.1968

[5] Nguyen M, Joshi SG. Carbapenem resistance in *Acinetobacter baumannii*, and their importance in hospital-acquired infections: A scientific review. Journal of Applied Microbiology. 2021;**131**(6):2715-2738. DOI: 10.1111/jam.15130

[6] Doughari HJ, Ndakidemi PA, Human IS, Benade S. The ecology, biology and pathogenesis of *Acinetobacter spp.*: An overview. Microbes and Environments. 2011;**26**(2):101-112. DOI: 10.1264/jsme2.me10179

[7] Ma C, McClean S. Mapping global prevalence of *Acinetobacter baumannii* and recent vaccine development to tackle it. Vaccine. 2021;**9**(6):1-25. DOI: 10.3390/vaccines9060570

[8] Qin J, Feng Y, Lü X, Zong Z. Precise species identification for *Acinetobacter*: A genome-based study with description of two novel *Acinetobacter species*. mSystems (American Society for Microbiology). 2021;**6**(3):e0023721. DOI: 10.1128/mSystems.00237-21

[9] Nemec A et al. Genotypic and phenotypic characterization of the *Acinetobacter calcoaceticus-Acinetobacter baumannii* complex with the proposal of *Acinetobacter pittii spp*. nov. (formerly *Acinetobacter* genomic species 3) and *Acinetobacter nosocomialis spp.* nov. (formerly *Acinetobacter* genomic species 13TU). Research in Microbiology. 2011;**162**(4):393-404. DOI: 10.1016/j.resmic.2011.02.006

[10] Dexter C, Murray GL, Paulsen IT, Peleg AY. Community-acquired *Acinetobacter baumannii*: Clinical characteristics, epidemiology and pathogenesis. Expert Review of Anti-Infective Therapy. 2015;**13**(5):567-573. DOI: 10.1586/14787210.2015.1025055

[11] Villalón P et al. Dynamics of a sporadic nosocomial *Acinetobacter calcoaceticus-Acinetobacter baumannii* complex population. Frontiers in Microbiology. 2019;**10**:593. DOI: 10.3389/fmicb.2019.00593

[12] Asif M, Alvi IA, Rehman SU. Insight into *Acinetobacter baumannii*: Pathogenesis, global resistance, mechanisms of resistance, treatment options, and alternative modalities. Infection and Drug Resistance. 2018;**11**:1249-1260. DOI: 10.2147/IDR.S166750

[13] Pendleton JN, Gorman SP, Gilmore BF. Clinical relevance of the ESKAPE pathogens. Expert Review of Anti-Infective Therapy. 2013;**11**(3):297-308. DOI: 10.1586/eri.13.12

[14] Scott P et al. An outbreak of multidrug-resistant *Acinetobacter*

baumannii-calcoaceticus complex infection in the US military health care system associated with military operations in Iraq. Clinical Infectious Diseases. 2007;**44**(12):1577-1584. DOI: 10.1086/518170

[15] Garcia Casallas JC, Robayo-Amortegui H, Corredor-Rozo Z, Carrasco-Márquez AM, Escobar-Perez J. Bacteremia by colistin-resistant *Acinetobacter baumannii* isolate: A case report. Journal of Medical Case Reports. 2019;**13**(1):141. DOI: 10.1186/s13256-019-2062-3

[16] Papathanakos G, Andrianopoulos I, Papathanasiou A, Priavali E, Koulenti D, Koulouras V. Colistin-resistant *Acinetobacter baumannii* bacteremia: A serious threat for critically ill patients. Microorganisms. 2020;**8**(2):1-12. DOI: 10.3390/microorganisms8020287

[17] Jung J, Park W. *Acinetobacter species* as model microorganisms in environmental microbiology: Current state and perspectives. Applied Microbiology and Biotechnology. 2015;**99**(6):2533-2548. DOI: 10.1007/s00253-015-6439-y

[18] Almasaudi SB. *Acinetobacter spp.* as nosocomial pathogens: Epidemiology and resistance features. Saudi Journal of Biological Sciences. 2018;**25**(3):586-596. DOI: 10.1016/j.sjbs.2016.02.009

[19] Chang WN, Lu CH, Huang CR, Chuang YC. Community-acquired *Acinetobacter* meningitis in adults. Infection. 2000;**28**(6):395-397. DOI: 10.1007/s150100070013

[20] Chang HC, Wei YF, Dijkshoorn L, Vaneechoutte M, Tang CT, Chang TC. Species-level identification of isolates of the *Acinetobacter calcoaceticus-Acinetobacter baumannii* complex by sequence analysis of the 16S-23S rRNA gene spacer region. Journal of Clinical Microbiology. 2005;**43**(4):1632-1639. DOI: 10.1128/JCM.43.4.1632-1639.2005

[21] Dolzani L, Tonin E, Lagatolla C, Prandin L, Monti-Bragadin C. Identification of *Acinetobacter* isolates in the *A. calcoaceticus-A. baumannii* complex by restriction analysis of the 16S-23S rRNA intergenic-spacer sequences. Journal of Clinical Microbiology. 1995;**33**(5):1108-1113. DOI: 10.1128/jcm.33.5.1108-1113.1995

[22] Ehrenstein B et al. *Acinetobacter species* identification by using tRNA spacer fingerprinting. Journal of Clinical Microbiology. 1996;**34**(10):2414-2420. DOI: 10.1128/jcm.34.10.2414-2420.1996

[23] Gerner-Smidt P. Ribotyping of the *Acinetobacter calcoaceticus-Acinetobacter baumannii* complex. Journal of Clinical Microbiology. 1992;**30**(10):2680-2685. DOI: 10.1128/jcm.30.10.2680-2685.1992

[24] Fournier PE et al. Comparative genomics of multidrug resistance in *Acinetobacter baumannii*. PLoS Genetics. 2006;**2**(1):e7. DOI: 10.1371/journal.pgen.0020007

[25] Leung WS, Chu CM, Tsang KY, Lo FH, Lo KF, Ho PL. Fulminant community-acquired *Acinetobacter baumannii* pneumonia as a distinct clinical syndrome. Chest. 2006;**129**(1):102-109. DOI: 10.1378/chest.129.1.102

[26] Anstey NM, Currie BJ, Hassell M, Palmer D, Dwyer B, Seifert H. Community-acquired bacteremic *Acinetobacter* pneumonia in tropical Australia is caused by diverse strains of *Acinetobacter baumannii*, with carriage in the throat in at-risk groups. Journal of Clinical Microbiology. 2002;**40**(2):685-686. DOI: 10.1128/JCM.40.2.685-686.2002

[27] Richards AM, Abu Kwaik Y, Lamont RJ. Code blue: *Acinetobacter baumannii*, a nosocomial pathogen with a role in the oral cavity. Molecular Oral Microbiology. 2015;**30**(1):2-15. DOI: 10.1111/omi.12072

[28] Dent LL, Marshall DR, Pratap S, Hulette RB. Multidrug resistant *Acinetobacter baumannii*: A descriptive study in a city hospital. BMC Infectious Diseases. 2010;**10**:196. DOI: 10.1186/1471-2334-10-196

[29] Rosenthal VD et al. International nosocomial infection control consortium (INICC) report, data summary of 45 countries for 2013-2018, adult and pediatric units, device-associated module. American Journal of Infection Control. 2021;**49**(10):1267-1274. DOI: 10.1016/j.ajic.2021.04.077

[30] Wendt C, Dietze B, Dietz E, Rüden H. Survival of *Acinetobacter baumannii* on dry surfaces. Journal of Clinical Microbiology. 1997;**35**(6):1394-1397. DOI: 10.1128/jcm.35.6.1394-1397.1997

[31] Castanheira M, Mendes RE, Gales AC. Global epidemiology and mechanisms of resistance of *Acinetobacter baumannii-calcoaceticus* complex. Clinical Infectious Diseases. 2023;**76**(Suppl. 2):S166-S178. DOI: 10.1093/cid/ciad109

[32] Doi Y, Murray GL, Peleg AY. *Acinetobacter baumannii*: Evolution of antimicrobial resistance-treatment options. Seminars in Respiratory and Critical Care Medicine. 2015;**36**(1):85-98. DOI: 10.1055/s-0034-1398388

[33] Cai Y, Chai D, Wang R, Liang B, Bai N. Colistin resistance of *Acinetobacter baumannii*: Clinical reports, mechanisms and antimicrobial strategies. The Journal of Antimicrobial Chemotherapy. 2012;**67**(7):1607-1615. DOI: 10.1093/jac/dks084

[34] Ko KS et al. High rates of resistance to colistin and polymyxin B in subgroups of *Acinetobacter baumannii* isolates from Korea. The Journal of Antimicrobial Chemotherapy. 2007;**60**(5):1163-1167. DOI: 10.1093/jac/dkm305

[35] Park GE et al. Bloodstream infections caused by *Acinetobacter species* with reduced susceptibility to tigecycline: Clinical features and risk factors. International Journal of Infectious Diseases. 2017;**62**:26-31. DOI: 10.1016/j.ijid.2017.06.023

[36] Gales AC, Seifert H, Gur D, Castanheira M, Jones RN, Sader HS. Antimicrobial susceptibility of *Acinetobacter baumannii-calcoaceticus* complex and *Stenotrophomonas malitophilia* clinical isolates: Results from the SENTRY antimicrobial surveillance programme (1997-2016). Open Forum Infectious Diseases. 2019;**6**(Suppl. 1): S34-S46. DOI: 10.1093/ofid/ofy293

[37] Seifert H, Blondeau J, Lucaßen K, Utt EA. Global update on the in vitro activity of tigecycline and comparators against isolates of *Acinetobacter baumannii* and rates of resistant phenotypes (2016-2018). Journal of Global Antimicrobial Resistance. 2022;**31**:82-89. DOI: 10.1016/j.jgar.2022.08.002

[38] Higgins PG, Dammhayn C, Hackel M, Seifert H. Global spread of carbapenem-resistant *Acinetobacter baumannii*. The Journal of Antimicrobial Chemotherapy. 2010;**65**(2):233-238. DOI: 10.1093/jac/dkp428

[39] Saeed NK, Kambal AM, El-Khizzi NA. Antimicrobial-resistant bacteria in a general intensive care unit in Saudi Arabia. Saudi Medical Journal. 2010;**31**(12):1341-1349

[40] Kamolvit W, Sidjabat HE, Paterson DL. Molecular epidemiology

and mechanisms of carbapenem resistance of *Acinetobacter spp.* in Asia and Oceania. Microbial Drug Resistance. 2015;**21**(4):424-434. DOI: 10.1089/mdr.2014.0234

[41] Al-Hassan L et al. Molecular epidemiology of carbapenem-resistant *Acinetobacter baumannii* from Khartoum State, Sudan. Frontiers in Microbiology. 2021;**12**:628736. DOI: 10.3389/fmicb.2021.628736

[42] Montrucchio G et al. The burden of carbapenem-resistant *Acinetobacter baumannii* in ICU COVID-19 patients: A regional experience. Journal of Clinical Medicine. 2022;**11**(17):1-11. DOI: 10.3390/jcm11175208

[43] Hartstein AI et al. Multiple intensive care unit outbreak of *Acinetobacter calcoaceticus subspecies anitratus* respiratory infection and colonization associated with contaminated, reusable ventilator circuits and resuscitation bags. The American Journal of Medicine. 1988;**85**(5):624-631. DOI: 10.1016/s0002-9343(88)80233-x

[44] Zarrilli R, Pournaras S, Giannouli M, Tsakris A. Global evolution of multidrug-resistant *Acinetobacter baumannii* clonal lineages. International Journal of Antimicrobial Agents. 2013;**41**(1):11-19. DOI: 10.1016/j.ijantimicag.2012.09.008

[45] Maragakis LL et al. An outbreak of multidrug-resistant *Acinetobacter baumannii* associated with pulsatile lavage wound treatment. JAMA. 2004;**292**(24):3006-3011. DOI: 10.1001/jama.292.24.3006

[46] Marchaim D et al. Surveillance cultures and duration of carriage of multidrug-resistant *Acinetobacter baumannii*. Journal of Clinical Microbiology. 2007;**45**(5):1551-1555. DOI: 10.1128/JCM.02424-06

[47] Naas T et al. VEB-1 extended-spectrum beta-lactamase-producing *Acinetobacter baumannii*, France. Emerging Infectious Diseases. 2006;**12**(8):1214-1222. DOI: 10.3201/eid1708.051547

[48] Coelho JM et al. Occurrence of carbapenem-resistant *Acinetobacter baumannii* clones at multiple hospitals in London and Southeast England. Journal of Clinical Microbiology. 2006;**44**(10):3623-3627. DOI: 10.1128/JCM.00699-06

[49] Odih EE et al. High genetic diversity of carbapenem-resistant *Acinetobacter baumannii* isolates recovered in Nigerian hospitals in 2016 to 2020. mSphere. 2023;**8**(3):e0009823. DOI: 10.1128/msphere.00098-23

[50] Odewale G, Adefioye OJ, Ojo J, Adewumi FA, Olowe OA. Multidrug resistance of *Acinetobacter baumannii* in Ladoke Akintola University Teaching Hospital, Osogbo, Nigeria. European Journal of Microbiology and Immunology. 2016;**6**(3):238-243. DOI: 10.1556/1886.2015.00018

[51] Olaitan AO, Berrazeg M, Fagade OE, Adelowo OO, Alli JA, Rolain JM. Emergence of multidrug-resistant *Acinetobacter baumannii* producing OXA-23 carbapenemase, Nigeria. International Journal of Infectious Diseases. 2013;**17**(6):e469-e470. DOI: 10.1016/j.ijid.2012.12.008

[52] Dizbay M, Tunccan OG, Sezer BE, Hizel K. Nosocomial imipenem-resistant *Acinetobacter baumannii* infections: Epidemiology and risk factors. Scandinavian Journal of Infectious Diseases. 2010;**42**(10):741-746. DOI: 10.3109/00365548.2010.489568

[53] Metan G, Sariguzel F, Sumerkan B. Factors influencing survival in patients with multi-drug-resistant *Acinetobacter* bacteraemia. European Journal of

Internal Medicine. 2009;**20**(5):540-544. DOI: 10.1016/j.ejim.2009.05.005

[54] Villegas MV, Hartstein AI. *Acinetobacter* outbreaks, 1977-2000. Infection Control and Hospital Epidemiology. 2003;**24**(4):284-295. DOI: 10.1086/502205

[55] Towner KJ. *Acinetobacter*: An old friend, but a new enemy. The Journal of Hospital Infection. 2009;**73**(4):355-363. DOI: 10.1016/j.jhin.2009.03.032

[56] Towner KJ. Clinical importance and antibiotic resistance of *Acinetobacter spp.* proceedings of a symposium held on 4-5 November 1996 at Eilat, Israel. Journal of Medical Microbiology. 1997;**46**(9):721-746. DOI: 10.1099/00222615-46-9-721

[57] McDonald LC, Banerjee SN, Jarvis WR. Seasonal variation of *Acinetobacter* infections: 1987-1996. Nosocomial Infections Surveillance System. Clinical Infectious Diseases. 1999;**29**(5):1133-1137. DOI: 10.1086/313441

[58] Weiner LM et al. Antimicrobial-resistant pathogens associated with healthcare-associated infections: Summary of data reported to the National Healthcare Safety Network at the Centers for Disease Control and Prevention, 2011-2014. Infection Control and Hospital Epidemiology. 2016;**37**(11):1288-1301. DOI: 10.1017/ice.2016.174

[59] Villers D et al. Nosocomial *Acinetobacter baumannii* infections: Microbiological and clinical epidemiology. Annals of Internal Medicine. 1998;**129**(3):182-189. DOI: 10.7326/0003-4819-129-3-199808010-00003

[60] Zhou H et al. Risk factors for acquisition and mortality of multidrug-resistant *Acinetobacter baumannii* bacteremia: A retrospective study from a Chinese hospital. Medicine. 2019;**98**(13):e14937. DOI: 10.1097/MD.0000000000014937

[61] Huang YC et al. Outbreak of *Acinetobacter baumannii* bacteremia in a neonatal intensive care unit: Clinical implications and genotyping analysis. The Pediatric Infectious Disease Journal. 2002;**21**(12):1105-1109. DOI: 10.1097/00006454-200212000-00004

[62] Mittal N et al. Outbreak of *Acinetobacter spp.* septicemia in a neonatal ICU. The Southeast Asian Journal of Tropical Medicine and Public Health. 2003;**34**(2):365-366

[63] Lolans K, Rice TW, Munoz-Price LS, Quinn JP. Multicity outbreak of carbapenem-resistant *Acinetobacter baumannii* isolates producing the carbapenemase OXA-40. Antimicrobial Agents and Chemotherapy. 2006;**50**(9):2941-2945. DOI: 10.1128/AAC.00116-06

[64] Siempos II, Vardakas KZ, Kyriakopoulos CE, Ntaidou TK, Falagas ME. Predictors of mortality in adult patients with ventilator-associated pneumonia: A meta-analysis. Shock. 2010;**33**(6):590-601. DOI: 10.1097/SHK.0b013e3181cc0418

[65] Eberle BM et al. The impact of *Acinetobacter baumannii* infections on outcome in trauma patients: A matched cohort study. Critical Care Medicine. 2010;**38**(11):2133-2138. DOI: 10.1097/CCM.0b013e3181f17af4

[66] Davis JS et al. A 16-year prospective study of community-onset bacteremic *Acinetobacter* pneumonia: Low mortality with appropriate initial empirical antibiotic protocols. Chest. 2014;**146**(4):1038-1045. DOI: 10.1378/chest.13-3065

[67] Anstey NM, Currie BJ, Withnall KM. Community-acquired *Acinetobacter*

pneumonia in the Northern Territory of Australia. Clinical Infectious Diseases. 1992;**14**(1):83-91. DOI: 10.1093/clinids/14.1.83

[68] Wang JT, McDonald LC, Chang SC, Ho M. Community-acquired *Acinetobacter baumannii* bacteremia in adult patients in Taiwan. Journal of Clinical Microbiology. 2002;**40**(4):1526-1529. DOI: 10.1128/JCM.40.4.1526-1529.2002

[69] Rafei R et al. Extrahuman epidemiology of *Acinetobacter baumannii* in Lebanon. Applied and Environmental Microbiology. 2015;**81**(7):2359-2367. DOI: 10.1128/AEM.03824-14

[70] Sheppard FR et al. The majority of US combat casualty soft-tissue wounds are not infected or colonized upon arrival or during treatment at a continental US military medical facility. American Journal of Surgery. 2010;**200**(4):489-495. DOI: 10.1016/j.amjsurg.2010.03.001

[71] Griffith ME, Lazarus DR, Mann PB, Boger JA, Hospenthal DR, Murray CK. *Acinetobacter* skin carriage among US army soldiers deployed in Iraq. Infection Control and Hospital Epidemiology. 2007;**28**(6):720-722. DOI: 10.1086/518966

[72] Hujer KM et al. Analysis of antibiotic resistance genes in multidrug-resistant *Acinetobacter spp.* isolates from military and civilian patients treated at the Walter Reed Army Medical Center. Antimicrobial Agents and Chemotherapy. 2006;**50**(12):4114-4123. DOI: 10.1128/AAC.00778-06

[73] Oncül O et al. Hospital-acquired infections following the 1999 Marmara earthquake. The Journal of Hospital Infection. 2002;**51**(1):47-51. DOI: 10.1053/jhin.2002.1205

[74] Neidell MJ et al. Costs of healthcare- and community-associated infections with antimicrobial-resistant versus antimicrobial-susceptible organisms. Clinical Infectious Diseases. 2012;**55**(6):807-815. DOI: 10.1093/cid/cis552

[75] Kim YJ, Kim SI, Hong KW, Kim YR, Park YJ, Kang MW. Risk factors for mortality in patients with carbapenem-resistant *Acinetobacter baumannii* bacteremia: Impact of appropriate antimicrobial therapy. Journal of Korean Medical Science. 2012;**27**(5):471-475. DOI: 10.3346/jkms.2012.27.5.471

[76] Kim SY et al. Risk factors for occurrence and 30-day mortality for carbapenem-resistant *Acinetobacter baumannii* bacteremia in an intensive care unit. Journal of Korean Medical Science. 2012;**27**(8):939-947. DOI: 10.3346/jkms.2012.27.8.939

[77] Antunes LC, Visca P, Towner KJ. *Acinetobacter baumannii*: Evolution of a global pathogen. Pathogens and Disease. 2014;**71**(3):292-301. DOI: 10.1111/2049-632X.12125

[78] Gordon NC, Wareham DW. Multidrug-resistant *Acinetobacter baumannii*: Mechanisms of virulence and resistance. International Journal of Antimicrobial Agents. 2010;**35**(3):219-226. DOI: 10.1016/j.ijantimicag.2009.10.024

[79] Garnacho-Montero J et al. *Acinetobacter baumannii* ventilator-associated pneumonia: Epidemiological and clinical findings. Intensive Care Medicine. 2005;**31**(5):649-655. DOI: 10.1007/s00134-005-2598-0

[80] Luna CM, Aruj PK. Nosocomial Acinetobacter pneumonia. Respirology. 2007;**12**(6):787-791. DOI: 10.1111/j.1440-1843.2007.01147.x

[81] Koulenti D, Tsigou E, Rello J. Nosocomial pneumonia in 27 ICUs in Europe: Perspectives from the EU-VAP/CAP study. European Journal of Clinical Microbiology & Infectious Diseases. 2017;**36**(11):1999-2006. DOI: 10.1007/s10096-016-2703-z

[82] Ong CW et al. Severe community-acquired *Acinetobacter baumannii* pneumonia: An emerging highly lethal infectious disease in the Asia-Pacific. Respirology. 2009;**14**(8):1200-1205. DOI: 10.1111/j.1440-1843.2009.01630.x

[83] Gaynes R, Edwards JR, System NNIS. Overview of nosocomial infections caused by gram-negative bacilli. Clinical Infectious Diseases. 2005;**41**(6):848-854. DOI: 10.1086/432803

[84] Wareham DW et al. Bloodstream infection due to *Acinetobacter spp*.: Epidemiology, risk factors and impact of multi-drug resistance. European Journal of Clinical Microbiology & Infectious Diseases. 2008;**27**(7):607-612. DOI: 10.1007/s10096-008-0473-y

[85] Wisplinghoff H, Edmond MB, Pfaller MA, Jones RN, Wenzel RP, Seifert H. Nosocomial bloodstream infections caused by *Acinetobacter species* in United States hospitals: Clinical features, molecular epidemiology, and antimicrobial susceptibility. Clinical Infectious Diseases. 2000;**31**(3):690-697. DOI: 10.1086/314040

[86] Cisneros JM et al. Bacteremia due to *Acinetobacter baumannii*: Epidemiology, clinical findings, and prognostic features. Clinical Infectious Diseases. 1996;**22**(6):1026-1032. DOI: 10.1093/clinids/22.6.1026

[87] Seifert H, Strate A, Pulverer G. Nosocomial bacteremia due to *Acinetobacter baumannii*. Clinical features, epidemiology, and predictors of mortality. Medicine. 1995;**74**(6): 340-349. DOI: 10.1097/00005792-199511000-00004

[88] Cisneros JM, Rodríguez-Baño J. Nosocomial bacteremia due to *Acinetobacter baumannii*: Epidemiology, clinical features and treatment. Clinical Microbiology and Infection. 2002;**8**(11):687-693. DOI: 10.1046/j.1469-0691.2002.00487.x

[89] Gómez J et al. Six-year prospective study of risk and prognostic factors in patients with nosocomial sepsis caused by *Acinetobacter baumannii*. European Journal of Clinical Microbiology & Infectious Diseases. 1999;**18**(5):358-361. DOI: 10.1007/pl00015019

[90] Guerrero DM et al. *Acinetobacter baumannii*-associated skin and soft tissue infections: Recognizing a broadening spectrum of disease. Surgical Infections. 2010;**11**(1):49-57. DOI: 10.1089/sur.2009.022

[91] Fleming ID et al. Modeling *Acinetobacter baumannii* wound infections: The critical role of iron. Journal of Trauma and Acute Care Surgery. 2017;**82**(3):557-565. DOI: 10.1097/TA.0000000000001338

[92] Siegman-Igra Y, Bar-Yosef S, Gorea A, Avram J. Nosocomial *Acinetobacter* meningitis secondary to invasive procedures: Report of 25 cases and review. Clinical Infectious Diseases. 1993;**17**(5):843-849. DOI: 10.1093/clinids/17.5.843

[93] Davis KA, Moran KA, McAllister CK, Gray PJ. Multidrug-resistant *Acinetobacter* extremity infections in soldiers. Emerging Infectious Diseases. 2005;**11**(8):1218-1224. DOI: 10.3201/1108.050103

[94] Wroblewska MM et al. Outbreak of nosocomial meningitis caused by

Acinetobacter baumannii in neurosurgical patients. The Journal of Hospital Infection. 2004;**57**(4):300-307. DOI: 10.1016/j.jhin.2004.04.009

[95] Grupper M, Sprecher H, Mashiach T, Finkelstein R. Attributable mortality of nosocomial *Acinetobacter* bacteremia. Infection Control and Hospital Epidemiology. 2007;**28**(3):293-298. DOI: 10.1086/512629

[96] García-Garmendia JL et al. Risk factors for *Acinetobacter baumannii* nosocomial bacteremia in critically ill patients: A cohort study. Clinical Infectious Diseases. 2001;**33**(7):939-946. DOI: 10.1086/322584

[97] Lee YT et al. Impact of appropriate antimicrobial therapy on mortality associated with *Acinetobacter baumannii* bacteremia: Relation to severity of infection. Clinical Infectious Diseases. 2012;**55**(2):209-215. DOI: 10.1093/cid/cis385

[98] Guo N, Xue W, Tang D, Ding J, Zhao B. Risk factors and outcomes of hospitalized patients with blood infections caused by multidrug-resistant *Acinetobacter baumannii* complex in a hospital of Northern China. American Journal of Infection Control. 2016;**44**(4):e37-e39. DOI: 10.1016/j.ajic.2015.11.019

[99] Valero C, Fariñas MC, García Palomo D, Mazarrasa JC, González Macías J. Endocarditis due to *Acinetobacter lwoffi* on native mitral valve. International Journal of Cardiology. 1999;**69**(1):97-99. DOI: 10.1016/s0167-5273(99)00006-6

[100] Rizos I, Tsiodras S, Papathanasiou S, Rigopoulos A, Barbetseas J, Stefanadis C. Prosthetic valve endocarditis due to *Acinetobacter spp.*: A rare case and literature review. The American Journal of the Medical Sciences. 2007;**333**(3):197-199. DOI: 10.1097/MAJ.0b013e31803193c4

[101] Malik AS. *Acinetobacter* endocarditis in children: A case report and review of the literature. Infection. 1995;**23**(5):306-308. DOI: 10.1007/BF01716293

[102] Fang G et al. Prosthetic valve endocarditis resulting from nosocomial bacteremia. A prospective, multicenter study. Annals of Internal Medicine. 1993;**119**(7 Pt 1):560-567. DOI: 10.7326/0003-4819-119-7_part_1-199310010-00003

[103] Kelkar R et al. Epidemic iatrogenic *Acinetobacter spp.* meningitis following administration of intrathecal methotrexate. The Journal of Hospital Infection. 1989;**14**(3):233-243. DOI: 10.1016/0195-6701(89)90040-6

[104] Levin AS, Levy CE, Manrique AE, Medeiros EA, Costa SF. Severe nosocomial infections with imipenem-resistant *Acinetobacter baumannii* treated with ampicillin/sulbactam. International Journal of Antimicrobial Agents. 2003;**21**(1):58-62. DOI: 10.1016/s0924-8579(02)00276-5

[105] Rodríguez Guardado A et al. Postsurgical meningitis caused by *Acinetobacter baumannii*: Study of 22 cases and review of the literature. Revista Clínica Española. 2001;**201**(9):497-500. DOI: 10.1016/s0014-2565(01)70895-8

[106] Chiang WC et al. Community-acquired bacteremic cellulitis caused by *Acinetobacter baumannii*. Journal of the Formosan Medical Association. 2003;**102**(9):650-652

[107] Bachmeyer C, Landgraf N, Cordier F, Lemaitre P, Blum L. *Acinetobacter baumanii* folliculitis in a patient with AIDS. Clinical and Experimental Dermatology. 2005;**30**(3):256-258. DOI: 10.1111/j.1365-2230.2005.01743.x

[108] Ng J, Gosbell IB, Kelly JA, Boyle MJ, Ferguson JK. Cure of multiresistant

Acinetobacter baumannii central nervous system infections with intraventricular or intrathecal colistin: Case series and literature review. The Journal of Antimicrobial Chemotherapy. 2006;**58**(5):1078-1081. DOI: 10.1093/jac/dkl347

[109] Adler BL, Krausz A, Friedman AJ. *Acinetobacter baumannii* emerging as a multidrug-resistant skin and soft-tissue pathogen: Parallels to methicillin-resistant *Staphylococcus aureus*. JAMA Dermatology. 2014;**150**(8):905-906. DOI: 10.1001/jamadermatol.2013.8855

[110] Sinha N, Niazi M, Lvovsky D. A fatal case of multidrug resistant *Acinetobacter* necrotizing fasciitis: The changing scary face of nosocomial infection. Case Reports in Infectious Diseases. 2014;**2014**:705279. DOI: 10.1155/2014/705279

[111] Turnidge J, Bell J, Biedenbach DJ, Jones RN. Pathogen occurrence and antimicrobial resistance trends among urinary tract infection isolates in the Asia-Western Pacific Region: Report from the SENTRY Antimicrobial Surveillance Program, 1998-1999. International Journal of Antimicrobial Agents. 2002;**20**(1):10-17. DOI: 10.1016/s0924-8579(02)00050-x

[112] Wang AG, Wu CC, Liu JH. Bacterial corneal ulcer: A multivariate study. Ophthalmologica. 1998;**212**(2):126-132. DOI: 10.1159/000027291

[113] Raut S et al. Trend and characteristics of *Acinetobacter baumannii* infections in patients attending Universal College of Medical Sciences Bhairahawa Western Nepal: A longitudinal study of 2018. Infection and Drug Resistance. 2020;**13**:1631-1641. DOI: 10.2147/IDR.S257851

[114] Kim SW et al. Serum resistance of *Acinetobacter baumannii* through the binding of factor H to outer membrane proteins. FEMS Microbiology Letters. 2009;**301**(2):224-231. DOI: 10.1111/j.1574-6968.2009.01820.x

[115] Gaddy JA, Actis LA. Regulation of *Acinetobacter baumannii* biofilm formation. Future Microbiology. 2009;**4**(3):273-278. DOI: 10.2217/fmb.09.5

[116] Jacobs AC et al. Inactivation of phospholipase D diminishes *Acinetobacter baumannii* pathogenesis. Infection and Immunity. 2010;**78**(5):1952-1962. DOI: 10.1128/IAI.00889-09

[117] Lee CR et al. Biology of *Acinetobacter baumannii:* Pathogenesis, antibiotic resistance mechanisms, and prospective treatment options. Frontiers in Cellular and Infection Microbiology. 2017;7:55. DOI: 10.3389/fcimb.2017.00055

Chapter 2

The Battle against Antibiotic Resistance: Novel Therapeutic Options for *Acinetobacter baumannii*

Amir Emami, Neda Pirbonyeh and Fatemeh Javanmardi

Abstract

Undoubtedly, *Acinetobacter baumannii* stands out as one of the most effective bacteria responsible for nosocomial infections within the healthcare system. Due to its multidrug-resistant nature and the frequency of outbreaks that it causes the treatment of infections caused by this bacterium is challenging, antimicrobial combination therapy has been utilized to treat multidrug resistance Gram-negatives when monotherapy is ineffective. In contrast to antibiotics or short peptides, which possess only the capacity to bind and regulate a specific target, antibodies exhibit supplementary properties attributed to their Fc region, including opsonophagocytic activity, the agglutination process, and activation of the complement system. The criticality of antibodies is exemplified in triggering immunity against *A. baumannii*, stimulating protective mechanisms, preventing bacterial attachment to epithelial cells, opsonization, and complement-dependent bacterial destruction. Given antibodies' significant role in humoral immunity, monoclonal antibodies (mAbs) may be generated to specifically bind to certain targets, thereby providing supplemental defense as a form of immunotherapy or passive immunization. Many encouraging tactics, ranging from phage therapy to immunotherapy, are being scrutinized for their efficacy in treating infectious diseases, thus shaping the future treatment landscape.

Keywords: antimicrobial peptides, bacteriophage therapy, drug repurposing, nanoparticles, MDR

1. Introduction

Bacterial infections are the leading cause of death worldwide. Although the discovery of antibiotics successfully controlled bacterial infections, overuse and misuse of antimicrobial agents exacerbated the selection of multidrug-resistant (MDR) organisms. *Acinetobacter baumannii* is a bacteria that increases infection and mortality in vulnerable patients due to its ability to escape from antibiotic treatments effectively.

The increasing prevalence of nosocomial *A. baumannii* infections can be largely attributed to the remarkable ability of *A. baumannii* to colonize and form biofilms.

Treatment options for these highly resistant pathogens are very limited. Because of this, clinicians are forced to resort to last-resort antibiotics, including colistin, which may induce nephrotoxicity, and select colistin-resistant *A. baumannii*.

To effectively treat and limit the spread of MDR *A. baumannii* (MDR-AB), a thorough understanding of the bacterial virulence factors and host-pathogen interactions is crucial. Therefore, before dealing with the new methods of treating *A. baumannii*, there is a need for a brief explanation to clarify the interactions between the host and the pathogen.

2. Virulence factors

A. baumannii, a highly antibiotic-resistant pathogen, possesses several virulence factors contributing to its pathogenesis and high mortality rates. Several recent studies have investigated virulence factors associated with the pathogenesis of *A. baumannii* and could thus serve as novel therapeutic targets.

These factors include the capsular polysaccharide (k-type), a major virulence factor [1]. The prevalent capsular types of *A. baumannii* include KL2, KL10, KL14, KL22, and KL52, with KL2 being associated with higher drug resistance and virulence [2].

Other virulence factors of *A. baumannii* include outer membrane proteins (Omps), lipopolysaccharide (LPS), capsular polysaccharide (CPS), phospholipase, nutrient-acquisition systems, efflux pumps, protein secretion systems, quorum sensing, and biofilm production (**Table 1**) [3]. Understanding these virulence factors is crucial for developing novel therapeutic targets and strategies to combat this multidrug-resistant pathogen [4, 5].

Virulence factor	Functions	Modulation
omps	Induce cell apoptosis, complement resistance, biofilm formation, cell invasion, and OMV biogenesis.	Unknown
CPS	Complement resistance and biofilm formation	Up-regulated upon antibiotic or ROS exposure
OMVs	Transferring OmpA and toxin delivery	Up-regulated upon antibiotic exposure
LPS	Membrane integrity, induce cell apoptosis, and antibiotic resistance	Loss during colistin resistance development
T6SS	Interspecies competition	Activate upon contact with competing bacteria
Micronutrient acquisition systems	Nutrient acquisition	Up-regulated under nutrient-deprived conditions
Type IV pili	Twitching motility	Up-regulated during growth in human serum
Bap	Biofilm formation	Up-regulated while growing under low iron conditions
Csu Pili	Biofilm formation	Antibiotic exposure

Table 1.
A. baumannii components, functions, and regulation conditions of them.

2.1 The mechanisms of *A. Baumannii* to promote self-survival

A. baumannii outer membrane proteins (Omps) play a versatile role in promoting bacterial survival. Omps in *A. baumannii* facilitate bacterial acclimatization to antibiotic- and host-induced stresses, aiding in immune evasion, stress tolerance, and resistance to antibiotics and antibacterial [6]. The ability of *A. baumannii* to adhere to abiotic surfaces and form biofilms, facilitated by Omps, and helps the bacteria survive in harsh environmental conditions such as desiccation, nutrient deficiency, and antibiotic treatment [7]. Additionally, *A. baumannii* outer membrane vesicles (OMVs), which contain Omps, contribute to the delivery of virulence factors to host cells, enhancing bacterial survival, nutrient acquisition, biofilm formation, and pathogenesis. *A. baumannii* strains that produce more abundant Omps, such as MDR-AB, exhibit more powerful cytotoxicity, stronger innate immune responses, and contain more virulence factors, potentially leading to worse outcomes [8].

LPS is the main component on the extracellular membrane of Gram-negative bacteria [9]. Mutations in the lipid A biosynthetic pathway can lead to changes in the structure of LPS in *A. baumannii*, resulting in colistin resistance. LPS-deficient *A. baumannii* strains show altered activation of the host innate immune inflammatory response, indicating the importance of LPS in interacting with host immune system. In addition, LPS-deficient *A. baumannii* can have alterations in their lipid A composition, such as the addition of phosphoethanolamine (pEtN) and galactosamine (GalN), which can affect the binding affinity of colistin. Loss of LPS in *A. baumannii* can lead to the upregulation of lipoproteins and the accumulation of the capsular polysaccharide poly-β-1,6-N-acetylglucosamine as compensatory mechanisms for membrane stabilization.

Capsular polysaccharides (CPS) in *A. baumannii* play a crucial role in bacterial virulence and survival [10]. The CPS structures in *A. baumannii* are diverse and can vary between strains [5]. These CPS structures often include rare sugars and branched oligosaccharide repeating units. The CPS biosynthesis gene encodes glycosyltransferases that are responsible for the synthesis of CPS structures. The presence of specific CPS structures, such as KL2, has been associated with antibiotic resistance and clinical outcomes in *A. baumannii* infections. Understanding the CPS structures and the genetics involved in their synthesis is important for developing targeted treatment strategies against *A. baumannii* infections [11].

Phospholipase functions of *A. baumannii* play a crucial role in promoting bacterial survival. Multiple studies have identified phospholipases as virulence factors that contribute to the pathogenicity of *A. baumannii*. The phospholipases are involved in the growth of phosphatidyl choline as a carbon source. These phospholipases involve various processes, such as hemolytic and cytolytic activities [12]. The phospholipases enable *A. baumannii* to adapt to different host niches and environments, enhance resistance to antimicrobial peptides, and facilitate the invasion of host cells [13].

Nutrient acquisition systems are often crucial for pathogen growth and survival during infection and represent attractive therapeutic targets. The pathogen utilizes various mechanisms to acquire essential nutrients from the host, such as heme and zinc.

A. baumannii has metal homeostatic systems that regulate the levels of essential nutrient metals in bacteria, particularly iron and zinc, that are important for colonizing different tissues and growth within vertebrates [14]. These systems, such as siderophores, heme uptake systems, and zinc uptake systems, enable the bacteria to overcome host-imposed zinc limitation by aiding in zinc uptake into the cells. The hemO locus, including the heme-degrading enzyme and scavenger, is required for high-affinity heme acquisition from host hemoglobin and serum albumin [15]. Additionally, *A. baumannii* possesses a Zn uptake (Znu) system consisting of an inner membrane ABC transporter and an outer membrane TonB-dependent receptor, which allows the pathogen to overcome host-imposed Zn limitation [16]. The TonB-dependent receptor HphR is an important component of the heme uptake system in *A. baumannii* and is involved in iron acquisition and cellular processes contributing to virulence [17].

Efflux pumps, such as the RND-type efflux pumps AdeABC and AdeIJK, contribute to resistance against antibiotics and biocides [18]. They are involved in extruding hazardous substances, including antibiotics, from within the bacterial cells [19]. The overexpression of these efflux pumps, particularly AdeABC, has been found to enhance the survival of *A. baumannii* when exposed to residual concentrations of biocides [20]. Additionally, efflux pump genes, such as adeABC, have been associated with tigecycline resistance in *A. baumannii* [21]. Efflux pumps have broad substrate specificity and are widely distributed among bacterial species, making them a major contributor to multidrug resistance in *A. baumannii* [22].

Secretion systems have recently been demonstrated to be involved in the pathogenic process, and five types of secretion systems out of the currently known six from Gram-negative bacteria have been found in *A. baumannii*. They can promote the bacteria's fitness and pathogenesis by releasing various effectors. Additionally, antibiotic resistance is found to be related to some types of secretion systems [23]. The type VI secretion system (T6SS) is one such system found in *A. baumannii*, which is involved in bacterial competition and the delivery of toxic effector proteins [24, 25]. The T6SS in *A. baumannii* is highly diverse, with significant diversity in the range of encoded T6SS VgrG and effector proteins. There are multiple VgrG genes in *A. baumannii* strains, with most strains encoding between two and four different VgrG proteins. T6SS structural components of *A. baumannii* are distinctive from other Gram-negative pathogens, as evidenced by the presence of the *Acinetobacter* genus-specific protein AsaA. The T6SS in *A. baumannii* is involved in bacterial competition and secretion of T6SS effectors, such as Hcp is associated, which acts as a virulence factor, transporter of effectors, and chaperone [26]. The putative T6SS effectors *in A. baumannii* have diverse functions, including peptidoglycan hydrolases, lipases, nucleases, and nucleic acid deaminases [27].

Quorum sensing (QS) in *A. baumannii* plays a crucial role in bacterial survival and pathogenicity [28]. The QS system coordinates the behavior of individual bacteria in a population by mediating the synthesis, secretion, and binding of auto-inducer signals. The deletion of the auto-inducer synthase gene abaI in *A. baumannii* resulted in a decrease in biofilm formation and pathogenicity [28]. Additionally, the QS system regulates important virulence-related phenotypes, such as surface-associated motility and biofilm formation [29]. The antibacterial peptide octopromycin inhibited biofilm formation and surface movements in *A. baumannii*, demonstrating its anti-quorum sensing activity [30]. Furthermore, the abaI/abaR QS system was found to affect growth characteristics, morphology, biofilm formation, resistance, motility, and virulence in *A. baumannii* [31].

Targeting virulence factors can be an effective strategy for combating *A. baumannii* and other multidrug-resistant bacteria. Additionally, the development of innovative strategies, such as using bacteriophages and antibiotics in combination, has shown increased efficacy in eradicating biofilms formed by antibiotic-resistant

A. baumannii strains. So, targeting virulence factors and biofilm formation can be an effective approach for designing drugs to combat multidrug-resistant bacteria.

3. Antimicrobial drug resistance and overcoming its problems

Antimicrobial drug resistance is one of the three major global threats to public health identified by the World Health Organization (WHO) in the twenty-first century [32]. *A. baumannii* is one of the main and most successful pathogens responsible for hospital-acquired infections in the modern healthcare system, associated with high mortality rates [33]. According to the reports of the WHO, about 80% of MDR or extensively drug-resistant (XDR) microbes have occurred due to the misuse and overuse of antibiotics, and these infections are associated with severe side effects [34]. Due to the prevalence of infections and outbreaks caused by *A. baumannii* drug resistance, few antibiotics are effective for treating infections caused by this pathogen [35]. Due to the spread of MDR bacteria and other resistant pathogens, there are limited treatment and prevention options, the failure of most antibiotics necessitates the search for better treatment options, and the need for alternative treatment options to treat these microbial pathogens (**Figure 1**).

3.1 Combined treatment

Antibiotics, such as colistin, carbapenems, and tigecycline, have been widely used to treat *A. baumannii* [36]. However, the emergence of this bacterium's multidrug-resistant strains has limited these drugs' effectiveness. Therefore, new treatment options like combination therapy are an urgent need. The idea behind combination therapy is to use two or more antibiotics to kill the bacteria, which works in different ways [37]. In fact, in this method, two antibiotics with other mechanisms of action are used, such as a β-lactam antibiotic and an aminoglycoside, or two antibiotics that have the same means of action but work on different targets, such as two different carbapenems [38].

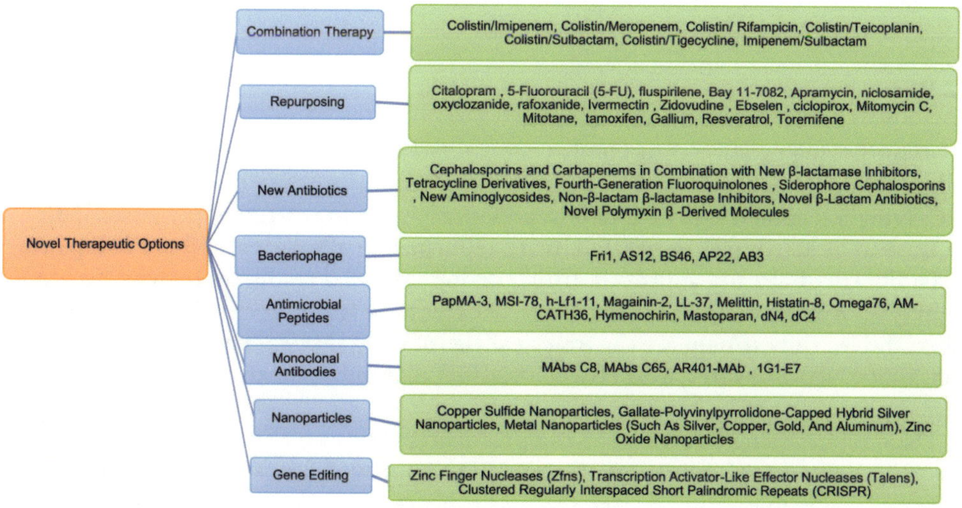

Figure 1.
Newer approaches to tackle of MDR A. baumannii.

There are several notable advantages to using combination therapy to treat *A. baumannii*. One of these advantages is increasing the effectiveness of treatment [39]. Bacteria are attacked from different angles, making it more difficult for bacteria to develop resistance using multiple drugs. Another advantage is that combination therapy can reduce the risk of treatment failure [40]. Since *A. baumannii* is very resistant to antibiotics, using one antibiotic may not be effective in treating the infection, using multiple antibiotics reduces the chance of treatment failure, using lower doses of each antibiotic can reduce the risk of side effects in patients.

For example, pairing β-lactam antibiotics with β-lactamase inhibitors has proven effective in combating resistant strains. Additionally, combining antibiotics with different mechanisms of action can target multiple bacterial pathways, increasing treatment efficacy [41].

Combination therapy, including colistin/imipenem, colistin/meropenem, colistin/rifampicin, colistin/teicoplanin, colistin/sulbactam, colistin/tigecycline, and imipenem/sulbactam has been widely studied [42].

Combination therapy has been explored as a potential treatment option for *A. baumannii* infections. Various combinations have been studied, including polymyxins, rifampicin, fosfomycin, sulbactam, and avibactam. Polymyxin-based combinations, such as with cell-wall acting agents, rifamycins, and fosfomycin, have been extensively studied [43, 44]. Berberine hydrochloride (BBH) has shown synergistic effects with antibiotics against MDR *A. baumannii* (MDR-AB), including tigecycline, sulbactam, meropenem, and ciprofloxacin [45]. High-dose sulbactam, combined with additional antibacterial agents, including colistin, has shown promise in treating MDR-AB or XDR *A. baumannii* (XDR-AB) infections [46]. Fosfomycin has also been explored as a potential component of combination therapy against carbapenem-resistant *A. baumannii* (CR-AB) conditions [47, 48].

The combination of colistin and tigecycline is effective in the treatment of *A. baumannii* infection that causes pneumonia by ventilator, and the combined therapy of colistin, meropenem, and ampicillin-sulbactam in *A. baumannii* infection in patients. It was effective in treating blood malignancy [49–51]. The combined treatment of colistin/rifampicin and ampicillin/sulbactam/carbapenem combination therapy is effective for the treatment of *A. baumannii* MDR bacteria, causing carbapenem-resistant skin and soft tissue infections [52].

3.2 Repurposing

Repurposing existing drugs is also considered a strategy for treating MDR bacterial infections. Repurposing drugs, drug repositioning, or therapeutic switching is like giving a second life to medication previously used for different purposes [53].

Instead of starting from scratch, drug repurposing allows researchers to tap into a vast library of already approved drugs, saving time and resources in drug development [54]. Moreover, low risk of failure, shorter time frame cycles, high success rates, and less investment are the practicalities of drug repurposing. These drugs have undergone rigorous safety and efficacy testing, making them attractive candidates for new applications [55].

Drug repurposing has emerged as a promising approach to combating drug-resistant *A. baumannii* infections. Several FDA-approved drugs have shown potential for repurposing in treating *A. baumannii*. Etoposide and genistein inhibit the synthesis of polyphosphates, a virulence factor in *A. baumannii* (**Table 2**) [56–58]. 5-fluorouracil (5-FU), fluspirilene, and Bay 11–7082 were identified as drugs that resensitize

	Compound	Activity-alone or in combination with	Approved use or known as
Central Nervous System	Citalopram	Polymyxin B	Antidepressant
	Fluspirilene	Colistin	Antipsychotic
Infectiology	Apramycin	Alone	Antibacterial
	Niclosamide	Colistin	Anti-helminthic
	Oxyclozanide	Alone	Anti-helminthic
	Rafoxanide	Alone	Anti-helminthic
	Ivermectin	Alone	Anti-parasitic
	Zidovudine	Alone	Antiretroviral
	Ciclopirox	Alone	Antifungal
Metabolism	Ebselen	Alone	Anti-inflammatory
	Bay 11–7082	Colistin	Anti-inflammatory
Natural Compound	Resveratrol	Colistin	Stilbene
Oncology	Mitomycin C	Alone	Anti-tumor
	Tamoxifen	Alone	Breast cancer
	5-Fluorouracil (5-FU)	Zidovudine	Antineoplastic (Colon Cancer)
	Mitotane	Polymyxin B	Antineoplastic
	Gallium	Alone	Antineoplastic
	Toremifene	Alone	Breast cancer

Table 2.
Relevant repurposing reports for MDR-AB.

A. baumannii to azithromycin and colistin in combination [59, 60]. Erythromycin, levamisole, chloroquine, and propranolol inhibit quorum sensing and virulence factors in *A. baumannii* [61, 62]. Tyrothricin, typically active against Gram-positive bacteria, exhibited antimicrobial activity against drug-resistant *A. baumannii* [63, 64].

Apramycin, Niclosamide, Oxyclozanide, Rafoxanide, and Ciclopirox are antibacterial, antifungal, and anthelmintic agents that have a therapeutic effect on MDR-AB.

Apramycin is an aminoglycoside approved for veterinary use. Apramycin can potentially be used against highly drug-resistant pathogens [65]. Niclosamide is an anthelmintic drug that has been commercially available in some countries since the 1960s. Niclosamide is usually administered orally and is well absorbed by the intestinal mucosa. High doses of this drug are associated with serious side effects. This drug has recently been suggested to treat other diseases, such as cancer [66].

Niclosamide alone has no antibacterial activity against *A. baumannii*, but a synergistic interaction between Niclosamide and colistin has been observed against CR-AB. This drug interacts with colistin-resistant strains negatively charged outer membrane, leading to a synergistic effect with colistin.

Oxyclozanide is used in veterinary medicine to treat fluke infections in ruminants. Oxyclozanide enhances the effect of colistin on colistin-sensitive and resistant isolates of *A. baumannii*, *P. aeruginosa*, and *K. pneumoniae*. This effect may be due to disrupting of the bacterial cell envelope [67].

Rafoxanide at a suitable dose had histidine kinase-antagonistic activities, which disrupted the abilities of MDR bacterial and fungal cells to adapt to stress conditions [68].

Ciclopirox, an antifungal drug, has bacteriostatic activity against *E. coli, K. pneumonia,* and *A. baumannii* of MDR strains [69]. The ciclopirox mechanism of action is the effect on the galactose and LPS salvage pathways [70].

Mitomycin C, tamoxifen, 5-FU, mitotane, and gallium include anti-tumor and antineoplastic drug agents that have a therapeutic effect on MDR-AB. The anticancer drug Mitomycin C can kill *A. baumannii* exponential-phase, stationary-phase, and biofilm cells [71].

The tamoxifen metabolites were active against MDR Gram-negative bacilli and might be potential antimicrobial agents to treat infections by these pathogens [72]. Colistin combination therapy with selective estrogen receptor modulators (SERM) as tamoxifen, raloxifene, and toremifen also exhibited good activity against polymyxin-resistant *P. aeruginosa*, *K. pneumonia*, and *A. baumannii* [73].

5-FU, another anticancer drug, despite the overall safety of 5-FU, is toxic in some cases, with toxicities including gastrointestinal (e.g., diarrhea, nausea, vomiting, mucositis/stomatitis, anorexia), hematological (e.g., neutropenia, thrombocytopenia, anemia), and dermal (e.g., hand-foot syndrome) symptoms [74]. The combination of 5-FU with azithromycin was effective against CR-AB; this combination, possibly reducing 5-FU toxicity, has also been found to inhibit the growth of bacterial pathogens and reduce the production of virulence factors [66].

Mitotane, an antineoplastic agent approved for cancer treatment, acts with polymyxin B on carbapenem- or polymyxin-resistant GNB *in vitro*. These efflux pump inhibitors alone did not affect the bacteria, but their activity was restored when combined with an antibiotic [70].

Gallium's antibacterial activity dates back many years, but this drug was originally used as an anticancer agent. Due to its chemical similarity to iron, gallium inhibits the reactions or redox pathways of iron and the growth of bacteria [75]. Therefore, gallium compounds show broad-spectrum antibacterial activity and inhibit the growth of important bacterial pathogens such as *A. baumannii*, *P. aeruginosa*, *S. aureus*, *K. pneumoniae*, and *E. coli* [76].

3.3 New antibiotics

Acinetobacter is one of the ESKAPE pathogens known for their ability to escape commonly used antibacterial treatments. With the rise of antibiotic resistance and the limited efficacy of current therapeutic options, exploring novel antibiotics offers hope in combating *A. baumannii* infections. New antibiotics have been developed and tested for treating MDR bacterial strains. These include cephalosporins and carbapenems in combination with new β-lactamase inhibitors, tetracycline derivatives, fourth-generation fluoroquinolones, new combinations of β-lactam and β-lactamase inhibitors, siderophore cephalosporins, and new aminoglycosides that have been approved or are in clinical development [77, 78].

Novel siderophore cephalosporins antibiotics, such as cefiderocol (CFDC) [79] and GT-1 (LCB10–0200), have shown promise for the treatment of *A. baumannii* infections. CFDC demonstrates strong activity against MDR-AB isolates with lower minimum inhibitory concentration (MIC) values than other Gram-negative agents [80]. GT-1, combined with a β-lactamase inhibitor GT-055, has shown efficacy against many multidrug-resistant pathogens, including *A. baumannii* [81]. These novel siderophore cephalosporins utilize a "trojan-horse approach" to evade resistance mechanisms in Gram-negative bacteria [82]. However, available clinical data for cefiderocol are conflicting, leaving infectious disease specialists uncertain about its optimal use in clinical practice.

New tetracycline antibiotics, such as eravacycline and TP-6076, have shown promise for treating Acinetob*acter* infections. Eravacycline has demonstrated higher potency than tigecycline and has been effective against XDR-AB *in vitro* [83]. It has also been found to have a low propensity to induce Clostridioides difficile infection (CDI) [84]. TP-6076, a fully synthetic fluorocycline, has shown greater activity than other tetracycline-class antimicrobials against CR-AB isolates [85]. These novel tetracyclines can be valuable additions to the limited armamentarium of drugs targeting *Acinetobacter* [86].

Non-β-lactam β-lactamase inhibitors antibiotics have shown promise for the treatment of *Acinetobacter* infections. The etx2514/sulbactam combination has demonstrated efficacy against MDR-AB isolates, including those producing class D β-lactamases [87]. The zidebactam/cefepime combination has shown *in vitro* activity against CR-AB [88]. Wck 4234/meropenem combination has exhibited broad-spectrum activity against MDR *Enterobacteriaceae*, including NDM, KPC, OXA, CTX-M, SHV, and TEM enzyme-producing isolates [89]. Ln-1-255/meropenem-imipenem combination has demonstrated decreased resistance rates against CR-AB isolates [90]. However, non-β-lactam β-lactamase inhibitors for treating *Acinetobacter* infections are still ongoing research and development. Further, clinical data is needed to support the efficacy of these inhibitors, and gaps still exist in the treatment of infections caused by MDR *Acinetobacter* spp.

Novel β-lactam antibiotics, such as AIC-499 and FSI-1671, combined with sulbactam have shown promise for treating *Acinetobacter* infections. AIC-499 is a member of the diazabicyclooctane class of β-lactamase inhibitors with broad-spectrum activity against Ambler class A, C, and D serine β-lactamases [87]. Sulbactam, a first-generation β-lactamase inhibitor, has limited action against *Acinetobacter* spp. due to susceptibility to cleavage by β-lactamases [91]. However, when combined with durlobactam, the activity of sulbactam is effectively restored against MDR *Acinetobacter* strains [92]. FSI-1671, in combination with sulbactam, has also shown enhanced antimicrobial activity against *A. baumannii* clinical strains in China, with cefoperazone-sulbactam as the most potent compound.

Novel polymyxin B-derived molecules, such as SPR741 and FADDI-287, have shown potential for treating *A. baumannii*. SPR741 has been found to potentiate several large scaffold antibiotics in Gram-negative pathogens by interacting predominantly with the outer membrane (OM) [93]. FADDI-287, on the other hand, has been shown to induce significant perturbation in glycerophospholipid metabolism and histidine degradation pathways, leading to synergistic bacterial killing in both polymyxin-susceptible and resistant *A. baumannii* [94]. These findings suggest that these novel polymyxin B-derived molecules can overcome resistance mechanisms and enhance the efficacy of antibiotics against *A. baumannii*.

A new aminoglycoside called apramycin (EBL-1003) has promising potential for treating *A. baumannii* infections. It has proven to have wide antibacterial action against *A. baumannii* strains resistant to various drugs, including standard-of-care aminoglycosides [95, 96]. Because of its distinct chemical makeup, apramycin can circumvent resistance mechanisms frequently present in clinical isolates that produce carbapenemase [97]. Apramycin is quickly bactericidal against *A. baumannii* according to *in vitro* experiments [98]. In a mouse lung infection model, apramycin was also discovered to have a high likelihood of target attainment and robust *in vivo* activity [99]. According to these findings, apramycin may be a potent therapeutic alternative for treating CR-AB lung infections linked to high mortality rates and few therapeutic options.

3.4 Bacteriophage

A. baumannii infections may be treated using bacteriophage therapy, particularly when there is MDR. Bacteriophages are viruses that can target and eradicate particular types of bacteria. The effectiveness and safety of phage therapy against *A. baumannii* infections have been demonstrated in several studies [91, 100, 101]. Other Gram-negative and Gram-positive bacteria, such as *S. aureus*, *P. aeruginosa*, *K. pneumoniae*, *Enterococcus*, and *Salmonella*, have been demonstrated to be sensitive to phage therapy. The effectiveness of phages can be assessed by evaluating their host range, adsorption rate, and growth curve from various sources, such as sewage or wastewater [102]. Bacteriophages, particularly lytic ones, have shown promise as anti-*A. baumannii* therapeutics [103]. Both options are using phages in monophage therapy, phage cocktails, or in conjunction with antibiotics [104]. Bacteriophages can be administered parenterally, orally, topically, or by inhalation [105]. Endolysins and depolymerases, two phage-derived enzymes, have also been investigated for use against *A. baumannii [106]*. Antimicrobial therapy can use specific lytic bacteriophages or enzymes generated from phages to treat infections brought on by strains of *A. baumannii* that are incredibly drug-resistant.

Bacteriophages have multiple mechanisms of action against *Acinetobacter*. Some phages produce depolymerase, such as the tail spike proteins of phages Fri1, AS12, BS46, and AP22, which specifically recognize and digest the capsular polysaccharide (CPS) of *A. baumannii*. Other phages can cause mutations in genes that alter the architecture of the bacterial envelope, leading to phage resistance but also increased sensitivity to antibiotics, such as colistin [107]. Additionally, phages can degrade biofilms formed by *A. baumannii*, as demonstrated by the bacteriophage AB3 and its endolysin LysAB3 [106]. Further, phage therapy can target specific mechanisms of antimicrobial resistance, such as efflux pumps, by using efflux pump inhibitors or phage steering [108]. So, phage therapy, including phage cocktails and combination therapy with antibiotics, as well as phage-derived enzymes, such as endolysins and depolymerase, shows promise in combating MDR-AB [109].

But, bacteriophage therapy for *Acinetobacter* infections has limitations that must be addressed. Firstly, there needs to be more reliable safety and efficacy data for phage therapy due to the heterogeneity in previously published studies [110]. Secondly, the systemic effects of phage therapy need to be better understood [111]. Thirdly, the optimal application protocol for phage therapy, including the route of administration, frequency of administration, treatment duration, and phage titer, is still [112]. Additionally, the concurrent ecological and evolutionary interplay between phages and host bacteria requires further research to utilize the potential of bacteriophage therapy [103].

3.5 Antimicrobial peptides

Antimicrobial peptides (AMPs) have been demonstrated to prevent the MDR bacteria *A. baumannii* from growing. AMPs with a strong antibacterial action against *A. baumannii* strains include PapMA-3, MSI-78, h-Lf1–11, magainin-2, and LL-37.

These peptides exhibit potential as antibacterial agents for treating bacteria that are resistant to antibiotics and have a broad spectrum of antimicrobial action [113].

Multiple mechanisms explain how AMPs work against *Acinetobacter.* The bacterial membrane may be damaged by AMPs, which will cause cell lysis [114, 115]. Additionally, they can infiltrate bacterial cells and engage with internal elements [116]. The mode of action of AMPs can be identified using experimental biophysical methods with model membranes and bacterial cells [117]. AMPs can cause the *Acinetobacter* membrane to become permeable [113]. The membrane's interaction with AMPs and lipids may result in soft supramolecular configurations, which can thin and lyse the membrane [118].

A hybrid peptide called PapMA-3 showed low cytotoxicity and strong bacterial selectivity against carbapenem-resistant bacteria [119]. *A. baumannii* was also significantly resistant to the antibacterial effects of MSI-78, h-Lf1–11, magainin-2, and LL-37 [120, 121]. The anti-*A. baumannii* potency of Melittin, Histatin-8, Omega76, AM-CATH36, Hymenochirin, (this peptide showed moderate activity against Gram-negative bacteria), and Mastoparan was the highest [113]. In animal models of *A. baumannii*-induced pneumonia, the AMP derivatives dN4 and dC4 have shown therapeutic effectiveness [119–122]. Furthermore, it has been demonstrated that larger doses of dN4 and dC4 can suppress and/or remove *Acinetobacter* biofilms [119].

A. baumannii, the cyclic peptide ZY4, exhibited little potential to induce resistance and outstanding efficacy against *A. baumannii*, including MDR strains [123]. According to these results, the antimicrobial peptides PapMA-3, MSI-78, h-Lf1–11, magainin-2, LL-37, Melittin, Histatin-8, Omega76, AM-CATH36, Hymenochirin, Mastoparan, dN4, and dC4 are efficient in treating *Acinetobacter* infections.

For the treatment of *Acinetobacter* infections, AMPs have limitations. Their toxicity and stability *in vivo* are two drawbacks that restrict their use [124]. Another drawback is their inherent limitations as peptides, such as stability, cytotoxicity, and bioavailability [125]. Natural AMPs are only useful for topical applications due to their pharmacological characteristics [126]. However, efforts are being made to get around these restrictions by creating novel AMPs and peptidomimetics by clever chemical changes [127]. Despite these drawbacks, AMPs hold potential as alternative therapies for focusing on bacterial infections, such as *Acinetobacter*, in both extracellular and intracellular contexts [113].

3.6 Monoclonal antibodies

Monoclonal antibodies (MAbs) are synthetic proteins, replicating the immune system's defense against pathogens like bacteria and cancer cells. These antibodies target molecules on the pathogen's surface known as antigens. The distinguishing quality of MAbs is their specificity, which enables them to recognize and bind to a certain target with extreme accuracy [128].

Heavy chains and light chains, two different protein chains, make up mAbs. These chains come together to form a Y-shaped structure. The antibody's variable region, found at the end of each Y-shaped arm and binds to the particular antigen [129]. On the other hand, the antibody's constant region controls its effector actions, such as triggering the immune system or obstructing the pathogen's activity [130, 131].

MAbs can use different pathways to exert their therapeutic effects. The process of neutralization, in which antibodies bind to the pathogen and stop it from infecting host cells, is a typical one. In order to enlist immune cells in the fight against the disease, antibodies can potentially trigger antibody-dependent cellular cytotoxicity (ADCC). Furthermore, MAbs can influence the immune system's response, enhancing the body's ability to eliminate the infection.

MAbs have shown promise as novel therapeutics for *Acinetobacter* infections. These antibodies targeting outer membrane protein A (OmpA) of *A. baumannii* have improved opsonophagocytic killing of the bacteria [132]. Another MAb targeting the capsule of *A. baumannii* has been found to enhance macrophage opsonophagocytosis and reduce pro-inflammatory cytokines, leading to improved survival in mouse models [133]. A second MAb, developed through hybridoma technology, has also been shown to enhance macrophage opsonophagocytosis and improve survival in murine models of *A. baumannii* infection alone and combination with antibiotics [134]. Furthermore, MAbs have the advantage of being a narrow spectrum, targeting only the pathogenic species and potentially avoiding microbiome disruption. Additionally, a human MAb targeting a DNABII epitope has demonstrated efficacy in disrupting biofilms formed by Gram-positive and Gram-negative bacteria, including *A. baumannii* [135]. These findings suggest that MAbs have the potential to be effective therapeutic options for *Acinetobacter* infections, either alone or in combination with antibiotics.

Modulation of pro- and anti-inflammatory cytokines, such as IL-1, IL-6, TNF, and IL-10, was necessary for MAb treatment to be effective [136]. The FDA has approved three antibacterial MAb medicines, and numerous others are undergoing clinical studies [137].

Treating *A. baumannii* infections has led to the development of the MAbs C8 and 65. In deadly bacteremic sepsis and aspiration pneumonia models of XDR-AB infection, MAb C8 improves opsonophagocytosis by focusing on the capsular carbohydrate on the bacterial surface [134]. On the other hand, MAb 65 is extremely powerful and effective while expanding the coverage of immunotherapeutic strains. Combined with antibiotics, such as colistin, it improves macrophage opsonophagocytosis and results [135]. These MAbs have demonstrated the ability to decrease cytokine production, blood bacterial density, and sepsis biomarkers, showing their therapeutic potential [132, 133]. Antibacterial MAb treatment works by regulating pro- and anti-inflammatory cytokines and improving germ clearance by opsonophagocytosis.

AR401-mAb is a monoclonal antibody developed for the treatment of *A. baumannii* infections. It is highly effective against a broad range of clinical isolates and has been shown to improve outcomes when combined with antibiotics [134]. AR401-MAb is synergistic with colistin, a commonly used antibiotic, further enhancing its protective effects [138]. These findings suggest that using MAbs, such as AR-401, in treating *A. baumannii* infections may effectively improve outcomes and reduce bacterial burden.

Another study produced MAbs against the outer membrane protein A (OmpA) of *A. baumannii*, with one MAb, 1G1-E7, showing high reactivity and opsonophagocytic killing activity [132].

3.7 Nanoparticles

Nanoparticles are exceedingly small particles, usually measuring between 1 and 100 nanometers. Various substances, including metals, metal oxides, lipids, and polymers, can be used to create them. Nanoparticles differ from their bulk counterparts in multiple ways due to their small size and frequently show improved reactivity and physical features.

Due to their unique characteristics, nanoparticles hold considerable potential for infection control. The ability of nanoparticles to carry antimicrobial drugs or to naturally have antimicrobial features makes them useful in fighting drug-resistant bacteria. They also interact with bacterial cells well due to their large surface area-to-volume ratio, strengthening their antimicrobial actions.

The bacterial cell membrane can be damaged by nanoparticle interaction, which results in cell death. They can pierce bacterial membranes, resulting in structural damage and cellular component release. This condition impairs the bacteria's capacity to continue performing essential tasks and ultimately results in their death. *Acinetobacter* is typically treated with nanoparticles by various mechanisms, including membrane damage, ROS production, efflux pump inhibition, and disruption of bacterial growth and biofilm formation. Copper sulfide nanoparticles, gallate-polyvinylpyrrolidone-capped hybrid silver nanoparticles, metal nanoparticles (such as silver, copper, gold, and aluminum), and zinc oxide nanoparticles are the specific types of nanoparticles utilized in treating *Acinetobacter*. These nanoparticles can potentially be potent therapeutic agents since they have demonstrated antibacterial activity against drug-resistant strains of *Acinetobacter*.

Nanoparticles, especially silver nanoparticles (AgNPs) and copper sulfide nanoparticles (cN16E-CuS), have shown promise in treating *A. baumannii* infections.

Silver nanoparticles (AgNPs) prevent the growth of drug-resistant strains by damaging bacterial membranes and producing reactive oxygen species (ROS). Biologically synthesized AgNPs also show efflux pump inhibitory activity, contributing to their antibacterial effect against MDR-AB. In addition, silver nanoparticles can induce apoptosis, inhibit the synthesis of new DNA in bacteria, and contribute to their antibacterial products. The antimicrobial activity of AgNPs is concentration-dependent and effective against extracellular and intracellular *A. baumannii*. Therefore, silver nanoparticles (AgNPs) have shown potential in treating *A. baumannii* infection. AgNPs showed good inhibitory activity against MDR-AB isolates, both alone and in combination with certain antibiotics. The combination of AgNPs with colistin, meropenem, or tigecycline significantly increased the sensitivity of MDR-AB to these antibiotics. In addition, silver nanoparticles inhibited the growth of *A. baumannii* and showed anti-biofilm activity, especially against weak biofilm producers. AgNPs have been found to interfere with the development of *A. baumannii* and disrupt biofilm formation, leading to a decrease in the expression of virulence and biofilm genes. Biogenic silver nanoparticles (Bio-AgNPs) synthesized by Fusarium oxysporum have also demonstrated antibacterial activity against CR-AB. Combined with polymyxin B, they showed synergistic effects, reducing the viable *A. baumannii* cells.

Another study reported using cationic antimicrobial lipid-stabilized copper sulfide nanoparticles (cN16E-CuS) for treating CR-AB. cN16E-CuS exhibited excellent antimicrobial activity against *A. baumannii*, producing excess reactive oxygen species and damaging bacterial membranes [139]. These findings suggest that nanoparticles, such as AgNPs and cN16E-CuS, can be used as alternative treatments for *A. baumannii* infections.

3.8 Gene editing

The DNA of creatures, including bacteria, can be changed using groundbreaking gene editing. It entails precise genetic manipulations such as adding, deleting, or changing particular genes. By concentrating on and interrupting the genes important for antibiotic resistance and other virulence factors, this method holds enormous potential for fighting bacterial infections.

Several gene editing tools have been developed to target bacterial pathogens, including *A. baumannii*. These tools include zinc finger nucleases (ZFNs) [140], transcription activator-like effector nucleases (TALENs) [141], and clustered regularly interspaced short palindromic repeats (CRISPR) [142] systems. Each tool offers unique advantages and can be tailored to target specific genes or regions within the bacterial genome.

CRISPR/Cas systems have shown potential as gene-editing tools for treating *A. baumannii* infections [143]. Genetic manipulation methods for studying *A. baumannii* pathogenesis and drug-resistance mechanisms are time-consuming and inefficient [144]. However, a detailed protocol for genetic manipulation in *A. baumannii*, including gene deletion, insertion, and point mutation, has been provided [145]. This protocol can aid in developing more innovative approaches to diagnosing and treating *A. baumannii* infections [146]. CRISPR/Cas systems can provide useful information about the functions of genes in *A. baumannii* and help identify potential targets for antimicrobials [140].

The Cas9 enzyme, which functions as molecular scissors, and a tiny RNA molecule known as a guide RNA, which points the Cas9 enzyme to the precise target spot in the bacterial genome, make up the CRISPR-Cas9 system.

Researchers have successfully applied the CRISPR-Cas9 system to target and edit the genes in *A. baumannii*. By designing appropriate guide RNA molecules, specific genes involved in antibiotic resistance or biofilm formation can be disrupted or modified, rendering the bacteria susceptible to existing antibiotics or impeding their ability to form biofilms.

CRISPR-Cas9 has been used for genetic manipulation in *A. baumannii* to study pathogenesis and drug-resistance mechanisms [145], which allowed for investigating drug-resistant mechanisms [147]. Additionally, a method for deleting drug-resistant genes in *A. baumannii* using CRISPR-Cas9 has been developed, providing a novel approach for preventing the spread of drug-resistant genes and treating drug-resistant bacteria [148].

3.9 Other

LpxC inhibitors have shown potential for the treatment of MDR-AB infections. Inhibiting LpxC, an enzyme involved in lipid biosynthesis, can reduce the toxicity of lipopolysaccharide (LPS) and enhance the efficacy of antibiotics [149, 150]. Compounds, such as LpxC-2 and LpxC-4, are synergistic with iron chelators (2,2′-bipyridyl and deferiprone) and gallium nitrate, significantly reducing bacterial counts.

The lipid A production is inhibited by LpxC inhibitors, such PF-5081090, which also increase cell permeability and improve resistance to a range of antibiotics such as rifampin, vancomycin, azithromycin, imipenem, and amikacin [72]. Additionally, LpxC inhibitors can prevent the activation of the Toll-like receptor 4 (TLR4) by *A. baumannii* LPS, which increases the opsonophagocytic death of the bacterium and decreases inflammation [150]. According to these results, LpxC inhibitors might be a different type of treatment for *A. baumannii* infections resistant to many drugs [151].

RX-P873, a novel antibiotic from the Pyrrolocytosine series, has shown high binding affinity for the bacterial ribosome and broad-spectrum antibiotic properties. It has demonstrated *in vitro* activity against MDR Gram-negative and Gram-positive strains of bacteria, including *A. baumannii*. In a study, RX-P873 was found to be highly active against *A. baumannii* isolates, with a MIC90 value of 1 μg/ml, which was two-fold more active than colistin and four-fold more active than tigecycline [152]. Additionally, a case report described the successful treatment of XDR-AB peritoneal dialysis-associated peritonitis with combination antibiotics, including intraperitoneal polymyxin B, without the need for catheter removal or switch to hemodialysis [153]. A study on RX-P873's activity against extracellular and intracellular forms of infection by *A. baumannii*, and other bacteria found that RX-P873 may be a useful alternative for disorders involving intracellular bacteria, especially Gram-negative species [154]. Therefore, RX-P873 shows potential as a treatment for *Acinetobacter* infections.

4. Conclusion

Novel therapeutic strategies for antimicrobial therapy of *Acinetobacter baumannii* include combination therapy, drug repurposing, novel antibiotics, bacteriophage therapy, antimicrobial peptides (AMPs), human monoclonal antibodies (Hu-mAbs), nanoparticles, and gene editing. These strategies aim to overcome drug resistance and improve the efficacy of treatment against extensively drug-resistant *Acinetobacter baumannii*.

Acknowledgements

We are grateful to the financial support by Burn Research Center, Shiraz University of Medical Sciences, Shiraz, Iran.

Conflict of interest

The authors declare that they have no conflict of interest.

Author details

Amir Emami[1*], Neda Pirbonyeh[1] and Fatemeh Javanmardi[2]

1 Microbiology Department, Burn & Wound Healing Research Center, Amir-Al-Momenin Burn Hospital, Shiraz University of Medical Sciences, Shiraz, Fars, Iran

2 Department of Biostatistics, School of Medicine, Shiraz University of Medical Sciences, Shiraz, Iran

*Address all correspondence to: emami.microbia@gmail.com

References

[1] Neto S, Vieira A, Oliveira H, Espiña B. Assessing *Acinetobacter baumannii* virulence and treatment with a bacteriophage using zebrafish embryos. The FASEB Journal. 2023;**37**(7):e23013. DOI: 10.1096/fj.202300385R

[2] Chen J, Li G, Wan F, Liu P, Du F, Shao Y, et al. Virulence characteristics and drug resistance of the prevalent capsule types in *Acinetobacter baumannii*. Microbial Drug Resistance. 2023;**29**(7):274-279. DOI: 10.1089/mdr.2022.0310

[3] Zhou JX, Feng DY, Li X, Zhu JX, Wu WB, Zhang TT. Advances in research on virulence factors of *Acinetobacter baumannii* and their potential as novel therapeutic targets. Journal of Applied Microbiology. 2023;**134**(2):lxac089. DOI: 10.1093/jambio/lxac089

[4] Shadan A, Pathak A, Ma Y, Pathania R, Singh RP. Deciphering the virulence factors, regulation, and immune response to *Acinetobacter baumannii* infection. Frontiers in Cellular and Infection Microbiology. 2023;**13**:1053968. DOI: 10.3389/fcimb.2023.1053968

[5] Dehbanipour R, Ghalavand Z. *Acinetobacter baumannii*: Pathogenesis, virulence factors, novel therapeutic options and mechanisms of resistance to antimicrobial agents with emphasis on tigecycline. Journal of Clinical Pharmacy and Therapeutics. 2022;**47**(11):1875-1884. DOI: 10.1111/jcpt.13787

[6] Uppalapati SR, Sett A, Pathania R. The outer membrane proteins OmpA, CarO, and OprD of *Acinetobacter baumannii* confer a two-pronged defense in facilitating its success as a potent human pathogen. Frontiers in Microbiology. 2020;**11**:589234. DOI: 10.3389/fmicb.2020.589234

[7] Li ZT, Zhang RL, Bi XG, Xu L, Fan M, Xie D, et al. Outer membrane vesicles isolated from two clinical *Acinetobacter baumannii* strains exhibit different toxicity and proteome characteristics. Microbial Pathogenesis. 2015;**81**:46-52. DOI: 10.1016/j.micpath.2015.03.009

[8] Zahn M, Bhamidimarri SP, Baslé A, Winterhalter M, van den Berg B. Structural insights into outer membrane permeability of *Acinetobacter baumannii*. Structure. 2016;**24**(2):221-231. DOI: 10.1016/j.str.2015.12.009

[9] Kamoshida G, Akaji T, Takemoto N, Suzuki Y, Sato Y, Kai D, et al. Lipopolysaccharide-deficient *Acinetobacter baumannii* due to colistin resistance is killed by neutrophil-produced lysozyme. Frontiers in Microbiology. 2020;**11**:573. DOI: 10.3389/fmicb.2020.00573

[10] Popova AV, Shneider MM, Arbatsky NP, Kasimova AA, Senchenkova SN, Shashkov AS, et al. Specific interaction of novel friunavirus phages encoding tailspike depolymerases with corresponding *Acinetobacter baumannii* capsular types. Journal of Virology. 2021;**95**(5):1-23. DOI: 10.1128/jvi.01714-20

[11] Yang JL, Yang CJ, Chuang YC, Sheng WH, Chen YC, Chang SC. Association of capsular polysaccharide locus 2 with prognosis of *Acinetobacter baumannii* bacteraemia. Emerging Microbes & Infections. 2022;**11**(1):83-90. DOI: 10.1080/22221751.2021.2011624

[12] Fiester SE, Arivett BA, Schmidt RE, Beckett AC, Ticak T, Carrier MV, et al.

Iron-regulated phospholipase C activity contributes to the cytolytic activity and virulence of *Acinetobacter baumannii*. PLoS One. 2016;**11**(11):e0167068. DOI: 10.1371/journal.pone.0167068

[13] Stahl J, Bergmann H, Göttig S, Ebersberger I, Averhoff B. *Acinetobacter baumannii* virulence is mediated by the concerted action of three phospholipases D. PLoS One. 2015;**10**(9):e0138360. DOI: 10.1371/journal.pone.0138360

[14] Mortensen BL, Skaar EP. The contribution of nutrient metal acquisition and metabolism to *Acinetobacter baumannii* survival within the host. Frontiers in Cellular and Infection Microbiology. 2013;**3**:95. DOI: 10.3389/fcimb.2013.00095

[15] Bateman TJ, Shah M, Ho TP, Shin HE, Pan C, Harris G, et al. A slam-dependent hemophore contributes to heme acquisition in the bacterial pathogen *Acinetobacter baumannii*. Nature Communications. 2021;**12**(1):6270. DOI: 10.1038/s41467-021-26545-9

[16] Hesse LE, Lonergan ZR, Beavers WN, Skaar EP. The *Acinetobacter baumannii* Znu system overcomes host-imposed nutrient zinc limitation. Infection and Immunity. 2019;**87**(12):e00746-e00719. DOI: 10.1128/iai.00746-19

[17] Zimbler DL, Arivett BA, Beckett AC, Menke SM, Actis LA. Functional features of TonB energy transduction systems of *Acinetobacter baumannii*. Infection and Immunity. 2013;**81**(9):3382-3394. DOI: 10.1128/iai.00540-13

[18] Meyer C, Lucaβen K, Gerson S, Xanthopoulou K, Wille T, Seifert H, et al. Contribution of RND-type efflux pumps in reduced susceptibility to biocides in *Acinetobacter baumannii*. Antibiotics (Basel). 2022;**11**(11):1635. DOI: 10.3390/antibiotics11111635

[19] Sharma S, Kaushik V, Kulshrestha M, Tiwari V. Different efflux pump systems in *Acinetobacter baumannii* and their role in multidrug resistance. Advances in Experimental Medicine and Biology. 2023;**1370**:155-168. DOI: 10.1007/5584_2023_771

[20] Mahapatra A, Pati A, Gupta K. Role of efflux pumps in reduced susceptibility to tigecycline among clinical isolates of *Acinetobacter baumannii*. International Journal of Applied & Basic Medical Research. 2022;**12**(4):243-248. DOI: 10.4103/ijabmr.ijabmr_219_22

[21] Verma P, Tiwari M, Tiwari V. Efflux pumps in multidrug-resistant *Acinetobacter baumannii*: Current status and challenges in the discovery of efflux pumps inhibitors. Microbial Pathogenesis. 2021;**152**:104766. DOI: 10.1016/j.micpath.2021.104766

[22] Su CC, Morgan CE, Kambakam S, Rajavel M, Scott H, Huang W, et al. Cryo-electron microscopy structure of an *Acinetobacter baumannii* multidrug efflux pump. MBio. 2019;**10**(4):e01295-e01219. DOI: 10.1128/mBio.01295-19

[23] Li P, Zhang S, Wang J, Al-Shamiri MM, Han B, Chen Y, et al. Uncovering the secretion systems of *Acinetobacter baumannii*: Structures and functions in pathogenicity and antibiotic resistance. Antibiotics (Basel). 2023;**12**(2):195. DOI: 10.3390/antibiotics12020195

[24] Li L, Wang YN, Jia HB, Wang P, Dong JF, Deng J, et al. The type VI secretion system protein AsaA in *Acinetobacter baumannii* is a periplasmic protein physically interacting with TssM and required for T6SS assembly. Scientific Reports. 2019;**9**(1):9438. DOI: 10.1038/s41598-019-45875-9

[25] Lewis JM, Deveson Lucas D, Harper M, Boyce JD.

Systematic identification and analysis of *Acinetobacter baumannii* type VI secretion system effector and immunity components. Frontiers in Microbiology. 2019;**10**:2440. DOI: 10.3389/fmicb.2019.02440

[26] Fitzsimons TC, Lewis JM, Wright A, Kleifeld O, Schittenhelm RB, Powell D, et al. Identification of novel *Acinetobacter baumannii* type VI secretion system antibacterial effector and immunity pairs. Infection and Immunity. 2018;**86**(8):e00297-e00218. DOI: 10.1128/iai.00297-18

[27] Ruiz FM, Santillana E, Spínola-Amilibia M, Torreira E, Culebras E, Romero A. Crystal structure of hcp from *Acinetobacter baumannii*: A component of the type VI secretion system. PLoS One. 2015;**10**(6):e0129691. DOI: 10.1371/journal.pone.0129691

[28] Xiong L, Yi F, Yu Q, Huang X, Ao K, Wang Y, et al. Transcriptomic analysis reveals the regulatory role of quorum sensing in the *Acinetobacter baumannii* ATCC 19606 via RNA-seq. BMC Microbiology. 2022;**22**(1):198. DOI: 10.1186/s12866-022-02612-z

[29] Rajapaksha DC, Edirisinghe SL, Nikapitiya C, Whang I, De Zoysa M. The antimicrobial peptide octopromycin suppresses biofilm formation and quorum sensing in *Acinetobacter baumannii*. Antibiotics (Basel). 2023;**12**(3):623. DOI: 10.3390/antibiotics12030623

[30] Mayer C, Muras A, Parga A, Romero M, Rumbo-Feal S, Poza M, et al. Quorum sensing as a target for controlling surface associated motility and biofilm formation in *Acinetobacter baumannii* ATCC(®) 17978(TM). Frontiers in Microbiology. 2020;**11**:565548. DOI: 10.3389/fmicb.2020.565548

[31] Sun X, Ni Z, Tang J, Ding Y, Wang X, Li F. The abaI/abaR quorum sensing system effects on pathogenicity in *Acinetobacter baumannii*. Frontiers in Microbiology. 2021;**12**:679241. DOI: 10.3389/fmicb.2021.679241

[32] Prestinaci F, Pezzotti P, Pantosti A. Antimicrobial resistance: A global multifaceted phenomenon. Pathogens and Global Health. 2015;**109**(7):309-318. DOI: 10.1179/2047773215y.0000000030

[33] Peleg AY, Seifert H, Paterson DL. *Acinetobacter baumannii*: Emergence of a successful pathogen. Clinical Microbiology Reviews. 2008;**21**(3):538-582. DOI: 10.1128/cmr.00058-07

[34] Uddin TM, Chakraborty AJ, Khusro A, Zidan BRM, Mitra S, Emran TB, et al. Antibiotic resistance in microbes: History, mechanisms, therapeutic strategies and future prospects. Journal of Infection and Public Health. 2021;**14**(12):1750-1766. DOI: 10.1016/j.jiph.2021.10.020

[35] Lin MF, Lan CY. Antimicrobial resistance in *Acinetobacter baumannii*: From bench to bedside. World Journal of Clinical Cases. 2014;**2**(12):787-814. DOI: 10.12998/wjcc.v2.i12.787

[36] Oikonomou O, Sarrou S, Papagiannitsis CC, Georgiadou S, Mantzarlis K, Zakynthinos E, et al. Rapid dissemination of colistin and carbapenem resistant *Acinetobacter baumannii* in Central Greece: Mechanisms of resistance, molecular identification and epidemiological data. BMC Infectious Diseases. 2015;**15**:559. DOI: 10.1186/s12879-015-1297-x

[37] Zhao X, Drlica K. Restricting the selection of antibiotic-resistant mutants: A general strategy derived from fluoroquinolone studies. Clinical

Infectious Diseases. 2001;**33**(Suppl. 3): S147-S156. DOI: 10.1086/321841

[38] Wang N, Luo J, Deng F, Huang Y, Zhou H. Antibiotic combination therapy: A strategy to overcome bacterial resistance to aminoglycoside antibiotics. Frontiers in Pharmacology. 2022;**13**:839808. DOI: 10.3389/fphar.2022.839808

[39] Adeniji OO, Nontongana N, Okoh JC, Okoh AI. The potential of antibiotics and nanomaterial combinations as therapeutic strategies in the management of multidrug-resistant infections: A review. International Journal of Molecular Sciences. 2022;**23**(23):15038. DOI: 10.3390/ijms232315038

[40] Dufour JF, Caussy C, Loomba R. Combination therapy for non-alcoholic steatohepatitis: Rationale, opportunities and challenges. Gut. 2020;**69**(10):1877-1884. DOI: 10.1136/gutjnl-2019-319104

[41] Narendrakumar L, Chakraborty M, Kumari S, Paul D. Das B.β-lactam potentiators to re-sensitize resistant pathogens: Discovery, development, clinical use and the way forward. Frontiers in Microbiology. 2022;**13**:1092556. DOI: 10.3389/fmicb.2022.1092556

[42] Liepiņš M, Krūmiņa A, Meistere I, Kosjkina D, Ķibilds J, Valciņa O, et al. Retrospective study of genetic diversity of *Acinetobacter baumannii*-resistant strains isolated from patients in Rīga east University Hospital in Latvia. Proceedings of the Latvian Academy of Sciences. 2021;**75**(2):142-148. DOI: 10.2478/prolas-2021-0022

[43] Scudeller L, Righi E, Chiamenti M, Bragantini D, Menchinelli G, Cattaneo P, et al. Systematic review and meta-analysis of *in vitro* efficacy of antibiotic combination therapy against carbapenem-resistant gram-negative bacilli. International Journal of Antimicrobial Agents. 2021;**57**(5):106344. DOI: 10.1016/j.ijantimicag.2021.106344

[44] Ardebili A, Izanloo A, Rastegar M. Polymyxin combination therapy for multidrug-resistant, extensively-drug resistant, and difficult-to-treat drug-resistant gram-negative infections: Is it superior to polymyxin monotherapy? Expert Review of Anti-Infective Therapy. 2023;**21**(4):387-429. DOI: 10.1080/14787210.2023.2184346

[45] Li X, Song Y, Wang L, Kang G, Wang P, Yin H, et al. A potential combination therapy of berberine hydrochloride with antibiotics against multidrug-resistant *Acinetobacter baumannii*. Frontiers in Cellular and Infection Microbiology. 2021;**11**:660431. DOI: 10.3389/fcimb.2021.660431

[46] Butler DA, Biagi M, Tan X, Qasmieh S, Bulman ZP, Wenzler E. Multidrug resistant *Acinetobacter baumannii*: Resistance by any other name would still be hard to treat. Current Infectious Disease Reports. 2019;**21**(12):46. DOI: 10.1007/s11908-019-0706-5

[47] Mazzitelli M, Gregori D, Sasset L, Trevenzoli M, Scaglione V, Lo Menzo S, et al. Cefiderocol-based versus Colistin-based regimens for severe carbapenem-resistant *Acinetobacter baumannii* infections: A propensity score-weighted, retrospective cohort study during the first two years of the COVID-19 pandemic. Microorganisms. 2023;**11**(4):984. DOI: 10.3390/microorganisms11040984

[48] Mohd Sazlly Lim S, Heffernan AJ, Roberts JA, Sime FB. Semi-mechanistic PK/PD modelling of fosfomycin and sulbactam combination against carbapenem-resistant *Acinetobacter*

baumannii. Antimicrobial Agents and Chemotherapy. 2023;**65**(5):e02472-e02420. DOI: 10.1128/aac.02472-20

[49] Falcone M, Tiseo G, Leonildi A, Della Sala L, Vecchione A, Barnini S, et al. Cefiderocol-compared to colistin-based regimens for the treatment of severe infections caused by carbapenem-resistant *Acinetobacter baumannii*. Antimicrobial Agents and Chemotherapy. 2022;**66**(5):e0214221. DOI: 10.1128/aac.02142-21

[50] Russo A, Bruni A, Gullì S, Borrazzo C, Quirino A, Lionello R, et al. Efficacy of cefiderocol-vs colistin-containing regimen for treatment of bacteraemic ventilator-associated pneumonia caused by carbapenem-resistant *Acinetobacter baumannii* in patients with COVID-19. International Journal of Antimicrobial Agents. 2023;**62**(1):106825. DOI: 10.1016/j.ijantimicag.2023.106825

[51] Liu J, Shu Y, Zhu F, Feng B, Zhang Z, Liu L, et al. Comparative efficacy and safety of combination therapy with high-dose sulbactam or colistin with additional antibacterial agents for multiple drug-resistant and extensively drug resistant *Acinetobacter baumannii* infections: A systematic review and network meta-analysis. Journal of Global Antimicrobial Resistance. 2021;**24**:136-147. DOI: 10.1016/j.jgar.2020.08.021

[52] Gautam D, Dolma KG, Khandelwal B, Mitsuwan W, Mahboob T, Pereira ML, et al. *Acinetobacter baumannii*: An overview of emerging multidrug-resistant pathogen. The Medical Journal of Malaysia. 2022;**77**(3):357-370

[53] Sonaye HV, Sheikh RY, Doifode CA. Drug repurposing: Iron in the fire for older drugs. Biomedicine & Pharmacotherapy. 2021;**141**:111638. DOI: 10.1016/j.biopha.2021.111638

[54] Choudhury C, Arul Murugan N, Priyakumar UD. Structure-based drug repurposing: Traditional and advanced AI/ML-aided methods. Drug Discovery Today. 2022;**27**(7):1847-1861. DOI: 10.1016/j.drudis.2022.03.006

[55] Krishnamurthy N, Grimshaw AA, Axson SA, Choe SH, Miller JE. Drug repurposing: A systematic review on root causes, barriers and facilitators. BMC Health Services Research. 2022;**22**(1):970. DOI: 10.1186/s12913-022-08272-z

[56] Gautam LK, Sharma P, Capalash N. Attenuation of *Acinetobacter baumannii* virulence by inhibition of polyphosphate kinase 1 with repurposed drugs. Microbiological Research. 2021;**242**:126627. DOI: 10.1016/j.micres.2020.126627

[57] Law SKK, Tan HS. The role of quorum sensing, biofilm formation, and iron acquisition as key virulence mechanisms in *Acinetobacter baumannii* and the corresponding anti-virulence strategies. Microbiological Research. 2022;**260**:127032. DOI: 10.1016/j.micres.2022.127032

[58] Terzioğlu E, Arslan M, Balaban BG, Çakar ZP. Microbial silver resistance mechanisms: Recent developments. World Journal of Microbiology and Biotechnology. 2022;**38**(9):158. DOI: 10.1007/s11274-022-03341-1

[59] Cheng YS, Sun W, Xu M, Shen M, Khraiwesh M, Sciotti RJ, et al. Repurposing screen identifies unconventional drugs with activity against multidrug resistant *Acinetobacter baumannii*. Frontiers in Cellular and Infection Microbiology. 2018;**8**:438. DOI: 10.3389/fcimb.2018.00438

[60] Ayon NJ. High-throughput screening of natural product and synthetic

molecule libraries for antibacterial drug discovery. Metabolites. 2023;**13**(5):625. DOI: 10.3390/metabo13050625

[61] Seleem NM, Abd El Latif HK, Shaldam MA, El-Ganiny A. Drugs with new lease of life as quorum sensing inhibitors: For combating MDR *Acinetobacter baumannii* infections. European Journal of Clinical Microbiology & Infectious Diseases. 2020;**39**(9):1687-1702. DOI: 10.1007/s10096-020-03882-z

[62] Zhong S, He S. Quorum sensing inhibition or quenching in *Acinetobacter baumannii*: The novel therapeutic strategies for new drug development. Frontiers in Microbiology. 2021;**12**:558003. DOI: 10.3389/fmicb.2021.558003

[63] Kundu R. Cationic amphiphilic peptides: Synthetic antimicrobial agents inspired by nature. Chem Med Chem. 2020;**15**(20):1887-1896. DOI: 10.1002/cmdc.202000301

[64] Han Y, Zhang M, Lai R, Zhang Z. Chemical modifications to increase the therapeutic potential of antimicrobial peptides. Peptides. 2021;**146**:170666. DOI: 10.1016/j.peptides.2021.170666

[65] Yong Y, Zhou Y, Liu K, Liu G, Wu L, Fang B. Exogenous citrulline and glutamine contribute to reverse the resistance of *Salmonella* to apramycin. Frontiers in Microbiology. 2021;**12**:759170. DOI: 10.3389/fmicb.2021.759170

[66] Gontijo AVL, Pereira SL, de Lacerda BH. Can drug repurposing be effective against carbapenem-resistant *Acinetobacter baumannii*? Current Microbiology. 2021;**79**(1):13. DOI: 10.1007/s00284-021-02693-5

[67] Ayerbe-Algaba R, Gil-Marqués ML, Miró-Canturri A, Parra-Millán R, Pachón-Ibáñez ME, Jiménez-Mejías ME, et al. The anthelmintic oxyclozanide restores the activity of colistin against colistin-resistant gram-negative bacilli. International Journal of Antimicrobial Agents. 2019;**54**(4):507-512. DOI: 10.1016/j.ijantimicag.2019.07.006

[68] Bendary MM, Abd El-Hamid MI, Abousaty AI, Elmanakhly AR, Alshareef WA, Mosbah RA, et al. Therapeutic switching of rafoxanide: A new approach to fighting drug-resistant bacteria and fungi. Microbiology Spectrum. 2023;**11**(4):e0267922. DOI: 10.1128/spectrum.02679-22

[69] Carlson-Banning KM, Chou A, Liu Z, Hamill RJ, Song Y, Zechiedrich L. Toward repurposing ciclopirox as an antibiotic against drug-resistant *Acinetobacter baumannii*, *Escherichia coli*, and *Klebsiella pneumoniae*. PLoS One. 2013;**8**(7):e69646. DOI: 10.1371/journal.pone.0069646

[70] Peyclit L, Baron SA, Rolain JM. Drug repurposing to fight colistin and carbapenem-resistant bacteria. Frontiers in Cellular and Infection Microbiology. 2019;**9**:193. DOI: 10.3389/fcimb.2019.00193

[71] Cruz-Muñiz MY, López-Jacome LE, Hernández-Durán M, Franco-Cendejas R, Licona-Limón P, Ramos-Balderas JL, et al. Repurposing the anticancer drug mitomycin C for the treatment of persistent *Acinetobacter baumannii* infections. International Journal of Antimicrobial Agents. 2017;**49**(1):88-92. DOI: 10.1016/j.ijantimicag.2016.08.022

[72] Miró-Canturri A, Vila-Domínguez A, Caretero-Ledesma M, Ayerbe-Algaba R, Pachón J, Jiménez-Mejías ME, et al. Repurposing of the tamoxifen metabolites to treat methicillin-resistant *Staphylococcus epidermidis* and

vancomycin-resistant *enterococcus faecalis* infections. Microbiology Spectrum. 2021;**9**(2):e0040321. DOI: 10.1128/Spectrum.00403-21

[73] Hussein M, Han ML, Zhu Y, Schneider-Futschik EK, Hu X, Zhou QT, et al. Mechanistic insights from global metabolomics studies into synergistic bactericidal effect of a polymyxin B combination with tamoxifen against cystic fibrosis MDR *Pseudomonas aeruginosa*. Computational and Structural Biotechnology Journal. 2018;**16**:587-599. DOI: 10.1016/j.csbj.2018.11.001

[74] Brutcher E, Christensen D, Hennessey Smith M, Koutlas JB, Sellers JB, Timmons T, et al. 5-fluorouracil and capecitabine: Assessment and treatment of uncommon early-onset severe toxicities associated with administration. Clinical Journal of Oncology Nursing. 2018;**22**(6):627-634. DOI: 10.1188/18.cjon.627-634

[75] Antunes LC, Imperi F, Minandri F, Visca P. In vitro and in vivo antimicrobial activities of gallium nitrate against multidrug-resistant *Acinetobacter baumannii*. Antimicrobial Agents and Chemotherapy. 2012;**56**(11):5961-5970. DOI: 10.1128/aac.01519-12

[76] Hijazi S, Visaggio D, Pirolo M, Frangipani E, Bernstein L, Visca P. Antimicrobial activity of gallium compounds on ESKAPE pathogens. Frontiers in Cellular and Infection Microbiology. 2018;**8**:316. DOI: 10.3389/fcimb.2018.00316

[77] Wagenlehner FME. New antibiotics for the treatment of urinary tract infections. Urologie. 2023;**62**(7):705-710. DOI: 10.1007/s00120-023-02121-5

[78] Terreni M, Taccani M, Pregnolato M. New antibiotics for multidrug-resistant bacterial strains: Latest research developments and future perspectives. Molecules. 2021;**26**(9):2671. DOI: 10.3390/molecules26092671

[79] Abdul-Mutakabbir JC, Nguyen L, Maassen PT, Stamper KC, Kebriaei R, Kaye KS, et al. In vitro antibacterial activity of cefiderocol against multidrug-resistant *Acinetobacter baumannii*. Antimicrobial Agents and Chemotherapy. 2021;**65**(9):e0264620. DOI: 10.1128/aac.02646-20

[80] Halasohoris SA, Scarff JM, Pysz LM, Lembirik S, Lemmon MM, Biek D, et al. *In vitro* and in vivo activity of GT-1, a novel siderophore cephalosporin, and GT-055, a broad-spectrum β-lactamase inhibitor, against biothreat and ESKAPE pathogens. Journal of Antibiotics (Tokyo). 2021;**74**(12):884-892. DOI: 10.1038/s41429-021-00472-9

[81] Abdul-Mutakabbir JC, Alosaimy S, Morrisette T, Kebriaei R, Rybak MJ. Cefiderocol: A novel siderophore cephalosporin against multidrug-resistant gram-negative pathogens. Pharmacotherapy. 2020;**40**(12):1228-1247. DOI: 10.1002/phar.2476

[82] Nguyen LP, Park CS, Pinto NA, Lee H, Seo HS, Vu TN, et al. *In vitro* activity of a novel siderophore-cephalosporin LCB10-0200 (GT-1), and LCB10-0200/avibactam, against carbapenem-resistant *Escherichia coli*, *Klebsiella pneumoniae*, *Acinetobacter baumannii*, and *Pseudomonas aeruginosa* strains at a tertiary hospital in Korea. Pharmaceuticals (Basel). 2021;**14**(4):370. DOI: 10.3390/ph14040370

[83] Yin T, Lai JJ, Huang WC, Kuo SC, Chiang TT, Yang YS. In vitro and *in vivo* comparison of eravacycline- and tigecycline-based combination therapies for tigecycline-resistant *Acinetobacter baumannii*. Journal of

Chemotherapy. 2022;**34**(3):166-172. DOI: 10.1080/1120009x.2021.2005755

[84] Buckley AM, Altringham J, Clark E, Bently K, Spittal W, Ewin D, et al. Eravacycline, a novel tetracycline derivative, does not induce clostridioides difficile infection in an in vitro human gut model. The Journal of Antimicrobial Chemotherapy. 2021;**76**(1):171-178. DOI: 10.1093/jac/dkaa386

[85] Van Hise N, Petrak RM, Skorodin NC, Fliegelman RM, Anderson M, Didwania V, et al. A real-world assessment of clinical outcomes and safety of eravacycline: A novel fluorocycline. Infectious Disease and Therapy. 2020;**9**(4):1017-1028. DOI: 10.1007/s40121-020-00351-0

[86] Seifert H, Stefanik D, Olesky M, Higgins PG. In vitro activity of the novel fluorocycline TP-6076 against carbapenem-resistant *Acinetobacter baumannii*. International Journal of Antimicrobial Agents. 2020;**55**(1):105829. DOI: 10.1016/j.ijantimicag.2019.10.010

[87] Shapiro AB, Moussa SH, McLeod SM, Durand-Réville T, Miller AA. Durlobactam, a new diazabicyclooctane β-lactamase inhibitor for the treatment of *Acinetobacter* infections in combination with sulbactam. Frontiers in Microbiology. 2021;**12**:709974. DOI: 10.3389/fmicb.2021.709974

[88] Yahav D, Giske CG, Grāmatniece A, Abodakpi H, Tam VH, Leibovici L. New β-lactam-β-lactamase inhibitor combinations. Clinical Microbiology Reviews. 2020;**34**(1):e00115-e00120. DOI: 10.1128/cmr.00115-20

[89] Barnes MD, Kumar V, Bethel CR, Moussa SH, O'Donnell J, Rutter JD, et al. Targeting multidrug-resistant *Acinetobacter* spp.: Sulbactam and the diazabicyclooctenone β-lactamase inhibitor ETX2514 as a novel therapeutic agent. MBio. 2019;**10**(2):e00159-e00119. DOI: 10.1128/mBio.00159-19

[90] Monogue ML, Giovagnoli S, Bissantz C, Zampaloni C, Nicolau DP. In vivo efficacy of meropenem with a novel non-β-lactam-β-lactamase inhibitor, nacubactam, against gram-negative organisms exhibiting various resistance mechanisms in a murine complicated urinary tract infection model. Antimicrobial Agents and Chemotherapy. 2018;**62**(9):e02596-e02517. DOI: 10.1128/aac.02596-17

[91] Wang L, Chen Y, Han R, Huang Z, Zhang X, Hu F, et al. Sulbactam enhances in vitro activity of β-lactam antibiotics against *Acinetobacter baumannii*. Infection and Drug Resistance. 2021;**14**:3971-3977. DOI: 10.2147/idr.s332160

[92] Lasarte-Monterrubio C, Vázquez-Ucha JC, Maneiro M, Arca-SuárezJ, AlonsoI, Guijarro-SánchezP, et al. Activity of Imipenem, meropenem, cefepime, and sulbactam in combination with the β-lactamase inhibitor LN-1-255 against *Acinetobacter* spp. Antibiotics (Basel). 2021;**10**(2):210. DOI: 10.3390/antibiotics10020210

[93] French S, Farha M, Ellis MJ, Sameer Z, Côté JP, Cotroneo N, et al. Potentiation of antibiotics against gram-negative bacteria by polymyxin B analogue SPR741 from unique perturbation of the outer membrane. ACS Infectious Diseases. 2020;**6**(6):1405-1412. DOI: 10.1021/acsinfecdis.9b00159

[94] Tran TB, Bergen PJ, Creek DJ, Velkov T, Li J. Synergistic killing of polymyxin B in combination with the antineoplastic drug mitotane against polymyxin-susceptible and -resistant *Acinetobacter baumannii*:

A metabolomic study. Frontiers in Pharmacology. 2018;**9**:359. DOI: 10.3389/fphar.2018.00359

[95] Becker K, Aranzana-Climent V, Cao S, Nilsson A, Shariatgorji R, Haldimann K, et al. Efficacy of EBL-1003 (apramycin) against *Acinetobacter baumannii* lung infections in mice. Clinical Microbiology and Infection. 2021;**27**(9):1315-1321. DOI: 10.1016/j.cmi.2020.12.004

[96] Juhas M, Widlake E, Teo J, Huseby DL, Tyrrell JM, Polikanov YS, et al. In vitro activity of apramycin against multidrug-, carbapenem- and aminoglycoside-resistant *Enterobacteriaceae* and *Acinetobacter baumannii*. The Journal of Antimicrobial Chemotherapy. 2019;**74**(4):944-952. DOI: 10.1093/jac/dky546

[97] Kang AD, Smith KP, Berg AH, Truelson KA, Eliopoulos GM, McCoy C, et al. Efficacy of Apramycin against multidrug-resistant *Acinetobacter baumannii* in the murine neutropenic thigh model. Antimicrobial Agents and Chemotherapy. 2018;**62**(4):e02585-e02517. DOI: 10.1128/aac.02585-17

[98] Smani Y, Pachón-Ibáñez ME, Pachón J. New molecules and adjuvants in the treatment of infections by *Acinetobacter baumannii*. Expert Opinion on Pharmacotherapy. 2016;**17**(9):1207-1214. DOI: 10.1080/14656566.2016.1176144

[99] Kang AD, Smith KP, Eliopoulos GM, Berg AH, McCoy C, Kirby JE. Invitro apramycin activity against multidrug-resistant *Acinetobacter baumannii* and *Pseudomonas aeruginosa*. Diagnostic Microbiology and Infectious Disease. 2017;**88**(2):188-191. DOI: 10.1016/j.diagmicrobio.2017.03.006

[100] Yan W, Banerjee P, Liu Y, Mi Z, Bai C, Hu H, et al. Development of thermosensitive hydrogel wound dressing containing *Acinetobacter baumannii* phage against wound infections. International Journal of Pharmaceutics. 2021;**602**:120508. DOI: 10.1016/j.ijpharm.2021.120508

[101] Broncano-LavadoA,Santamaría-CorralG, Esteban J, García-Quintanilla M. Advances in bacteriophage therapy against relevant multi drug-resistant pathogens. Antibiotics (Basel). 2021;**10**(6):672. DOI: 10.3390/antibiotics10060672

[102] Malik S, Nehra K, Rana JS. Bacteriophage cocktail and phage antibiotic synergism as promising alternatives to conventional antibiotics for the control of multi-drug-resistant uropathogenic *Escherichia coli*. Virus Research. 2021;**302**:198496. DOI: 10.1016/j.virusres.2021.198496

[103] Rai S, Kumar A. Bacteriophage therapeutics to confront multidrug-resistant *Acinetobacter baumannii*—A global health menace. Environmental Microbiology Reports. 2022;**14**(3):347-364. DOI: 10.1111/1758-2229.12988

[104] Düzgüneş N, Sessevmez M, Yildirim M. Bacteriophage therapy of bacterial infections: The rediscovered frontier. Pharmaceuticals (Basel). 2021;**14**(1):34. DOI: 10.3390/ph14010034

[105] Barber OW, Miramontes IM, Jain M, Ozer EA, Hartmann EM. The future of bacteriophage therapy will promote antimicrobial susceptibility. mSystems. 2021;**6**(4):e0021821. DOI: 10.1128/mSystems.00218-21

[106] Junghans S, Rojas SV, Skusa R, Püschel A, Grambow E, Kohlen J, et al. Bacteriophages for the treatment of graft infections in cardiovascular medicine.

Antibiotics (Basel). 2021;**10**(12):1446. DOI: 10.3390/antibiotics10121446

[107] Liu D, Van Belleghem JD, de Vries CR, Burgener E, Chen Q, Manasherob R, et al. The safety and toxicity of phage therapy: A review of animal and clinical studies. Viruses. 2021;**13**(7):1268. DOI: 10.3390/v13071268

[108] Jeon J, Park JH, Yong D. Efficacy of bacteriophage treatment against carbapenem-resistant *Acinetobacter baumannii* in galleria mellonella larvae and a mouse model of acute pneumonia. BMC Microbiology. 2019;**19**(1):70. DOI: 10.1186/s12866-019-1443-5

[109] Tan Y, Su J, Fu M, Zhang H, Zeng H. Recent advances in phage-based therapeutics for multi-drug resistant *Acinetobacter baumannii*. Bioengineering (Basel). 2022;**10**(1):35. DOI: 10.3390/bioengineering10010035

[110] Bhargava K, Nath G, Bhargava A, Aseri GK, Jain N. Phage therapeutics: From promises to practices and prospectives. Applied Microbiology and Biotechnology. 2021;**105**(24):9047-9067. DOI: 10.1007/s00253-021-11695-z

[111] Mu A, McDonald D, Jarmusch AK, Martino C, Brennan C, Bryant M, et al. Assessment of the microbiome during bacteriophage therapy in combination with systemic antibiotics to treat a case of staphylococcal device infection. Microbiome. 2021;**9**(1):92. DOI: 10.1186/s40168-021-01026-9

[112] Onsea J, Uyttebroek S, Chen B, Wagemans J, Lood C, Van Gerven L, et al. Bacteriophage therapy for difficult-to-treat infections: The implementation of a multidisciplinary phage task force (the PHAGEFORCE study protocol). Viruses. 2021;**13**(8):1543. DOI: 10.3390/v13081543

[113] Neshani A, Sedighian H, Mirhosseini SA, Ghazvini K, Zare H, Jahangiri A. Antimicrobial peptides as a promising treatment option against *Acinetobacter baumannii* infections. Microbial Pathogenesis. 2020;**146**:104238. DOI: 10.1016/j.micpath.2020.104238

[114] Luo Y, Song Y. Mechanism of antimicrobial peptides: Antimicrobial, anti-inflammatory and antibiofilm activities. International Journal of Molecular Sciences. 2021;**22**(21):11401. DOI: 10.3390/ijms222111401

[115] Zhang QY, Yan ZB, Meng YM, Hong XY, Shao G, Ma JJ, et al. Antimicrobial peptides: Mechanism of action, activity and clinical potential. Military Medical Research. 2021;**8**(1):48. DOI: 10.1186/s40779-021-00343-2

[116] Drayton M, Deisinger JP, Ludwig KC, Raheem N, Müller A, Schneider T, et al. Host defense peptides: Dual antimicrobial and immunomodulatory action. International Journal of Molecular Sciences. 2021;**22**(20):11172. DOI: 10.3390/ijms222011172

[117] Benfield AH, Henriques ST. Mode-of-action of antimicrobial peptides: Membrane disruption vs. intracellular mechanisms. Frontiers in Medical Technology. 2020;**2**:610997. DOI: 10.3389/fmedt.2020.610997

[118] Wadhwani P, Sekaran S, Strandberg E, Bürck J, Chugh A, Ulrich AS. Membrane interactions of Latarcins: Antimicrobial peptides from spider venom. International Journal of Molecular Sciences. 2021;**22**(18):10156. DOI: 10.3390/ijms221810156

[119] Choi J, Jang A, Yoon YK, Kim Y. Development of novel peptides for the antimicrobial combination therapy

against carbapenem-resistant *Acinetobacter baumannii* infection. Pharmaceutics. 2021;**13**(11):1800. DOI: 10.3390/pharmaceutics13111800

[120] Denardi LB, de Arruda TP, Weiblen C, Ianiski LB, Stibbe PC, Pinto SC, et al. In vitro activity of the antimicrobial peptides h-Lf1-11, MSI-78, LL-37, fengycin 2B, and magainin-2 against clinically important bacteria. Brazilian Journal of Microbiology. 2022;**53**(1):171-177. DOI: 10.1007/s42770-021-00645-6

[121] Ridyard KE, Overhage J. The potential of human peptide LL-37 as an antimicrobial and anti-biofilm agent. Antibiotics (Basel). 2021;**10**(6):650. DOI: 10.3390/antibiotics10060650

[122] Rima M, Rima M, Fajloun Z, Sabatier JM, Bechinger B, Naas T. Antimicrobialpeptides:Apotentalternative to antibiotics. Antibiotics (Basel). 2021;**10**(9):1095. DOI: 10.3390/antibiotics10091095

[123] Mwangi J, Yin Y, Wang G, Yang M, Li Y, Zhang Z, et al. The antimicrobial peptide ZY4 combats multidrug-resistant *Pseudomonas aeruginosa* and *Acinetobacter baumannii* infection. Proceedings of the National Academy of Sciences of the United States of America. 2019;**116**(52):26516-26522. DOI: 10.1073/pnas.1909585117

[124] Jung CJ, Liao YD, Hsu CC, Huang TY, Chuang YC, Chen JW, et al. Identification of potential therapeutic antimicrobial peptides against *Acinetobacter baumannii* in a mouse model of pneumonia. Scientific Reports. 2021;**11**(1):7318. DOI: 10.1038/s41598-021-86844-5

[125] Amiss AS, Henriques ST, Lawrence N. Antimicrobial peptides provide wider coverage for targeting drug-resistant bacterial pathogens. Peptide Science. 2022;**114**(2):e24246. DOI: 10.1002/pep2.24246

[126] Wang C, Hong T, Cui P, Wang J, Xia J. Antimicrobial peptides towards clinical application: Delivery and formulation. Advanced Drug Delivery Reviews. 2021;**175**:113818. DOI: 10.1016/j.addr.2021.05.028

[127] Ioannou P, Baliou S, Kofteridis DP. Antimicrobial peptides in infectious diseases and beyond—A narrative review. Life (Basel). 2023;**13**(8):1651. DOI: 10.3390/life13081651

[128] Horenstein AL, Chillemi A, Quarona V, Zito A, Mariani V, Faini AC, et al. Antibody mimicry, receptors and clinical applications. Human Antibodies. 2017;**25**(3-4):75-85. DOI: 10.3233/hab-160305

[129] Santoriello D, Markowitz GS. Heavy and light chains all at once: A new immunofluorescence technique to evaluate monoclonal immunoglobulin deposits. Kidney International. 2021;**100**(1):22-24. DOI: 10.1016/j.kint.2021.04.022

[130] Melani RD, Srzentić K, Gerbasi VR, McGee JP, Huguet R, Fornelli L, et al. Direct measurement of light and heavy antibody chains using ion mobility and middle-down mass spectrometry. MAbs. 2019;**11**(8):1351-1357. DOI: 10.1080/19420862.2019.1668226

[131] Posner J, Barrington P, Brier T, Datta-Mannan A. Monoclonal antibodies: Past, present and future. Handbook of Experimental Pharmacology. 2019;**260**:81-141. DOI: 10.1007/164_2019_323

[132] Yeganeh O, Shabani M, Pakzad P, Mosaffa N, Hashemi A. Production and characterization of novel monoclonal

antibodies against outer membrane protein a (OmpA) of *Acinetobacter baumannii*. Journal of Immunological Methods. 2021;**499**:113169. DOI: 10.1016/j.jim.2021.113169

[133] Nielsen TB, Yan J, Luna BM, Talyansky Y, Slarve M, Bonomo RA, et al. Monoclonal antibody requires immunomodulation for efficacy against *Acinetobacter baumannii* infection. The Journal of Infectious Diseases. 2021;**224**(12):2133-2147. DOI: 10.1093/infdis/jiab265

[134] Nielsen TB, Yan J, Slarve M, Lu P, Li R, Ruiz J, et al. Monoclonal antibody therapy against *Acinetobacter baumannii*. Infection and Immunity. 2021;**89**(10):e0016221. DOI: 10.1128/iai.00162-21

[135] Nielsen TB, Pantapalangkoor P, Luna BM, Bruhn KW, Yan J, Dekitani K, et al. Monoclonal antibody protects against *Acinetobacter baumannii* infection by enhancing bacterial clearance and evading sepsis. The Journal of Infectious Diseases. 2017;**216**(4):489-501. DOI: 10.1093/infdis/jix315

[136] Ouyang W. O'Garra a.IL-10 family cytokines IL-10 and IL-22: From basic science to clinical translation. Immunity. 2019;**50**(4):871-891. DOI: 10.1016/j.immuni.2019.03.020

[137] Wang-Lin SX, Balthasar JP. Pharmacokinetic and pharmacodynamic considerations for the use of monoclonal antibodies in the treatment of bacterial infections. Antibodies (Basel). 2018;**7**(1):5. DOI: 10.3390/antib7010005

[138] Skariyachan S, Taskeen N, Ganta M, Venkata KB. Recent perspectives on the virulent factors and treatment options for multidrug-resistant *Acinetobacter baumannii*. Critical Reviews in Microbiology. 2019;**45**(3):315-333. DOI: 10.1080/1040841x.2019.1600472

[139] Karnan Singaravelu D, Bala Subramaniyan S, Tharunya MP, Veerappan A. Antimicrobial lipid capped copper sulfide nanoparticles display enhanced bactericidal effect against carbapenem-resistant *Acinetobacter baumannii*. Materials Letters. 2022;**306**:130985. DOI: 10.1016/j.matlet.2021.130985

[140] Serajian S, Ahmadpour E, Oliveira SMR, Pereira ML, Heidarzadeh S. CRISPR-Cas technology: Emerging applications in clinical microbiology and infectious diseases. Pharmaceuticals (Basel). 2021;**14**(11):1171. DOI: 10.3390/ph14111171

[141] Thompson MK, Sobol RW, Prakash A. Exploiting DNA endonucleases to advance mechanisms of DNA repair. Biology (Basel). 2021;**10**(6):530. DOI: 10.3390/biology10060530

[142] Nargesi S, Kaboli S, Thekkiniath J, Heidari S, Keramati F, Seyedmousavi S, et al. Recent advances in genome editing tools in medical mycology research. Journal of Fungi (Basel). 2021;**7**(4):257. DOI: 10.3390/jof7040257

[143] Liu W, Li L, Jiang J, Wu M, Lin P. Applications and challenges of CRISPR-Cas gene-editing to disease treatment in clinics. Precision Clinical Medicine. 2021;**4**(3):179-191. DOI: 10.1093/pcmedi/pbab014

[144] Malech HL. Treatment by CRISPR-Cas9 gene editing—A proof of principle. The New England Journal of Medicine. 2021;**384**(3):286-287. DOI: 10.1056/NEJMe2034624

[145] Wang Y, Wang Z. Ji Q.CRISPR-Cas9-based genome editing and cytidine base editing in *Acinetobacter baumannii*. STAR Protocols. 2020;**1**(1):100025. DOI: 10.1016/j.xpro.2020.100025

[146] Zhang B. CRISPR/Cas gene therapy. Journal of Cellular Physiology. 2021;**236**(4):2459-2481. DOI: 10.1002/jcp.30064

[147] Wang Y, Wang Z, Chen Y, Hua X, Yu Y, Ji Q. A highly efficient CRISPR-Cas9-based genome engineering platform in *Acinetobacter baumannii* to understand the H(2)O(2)-sensing mechanism of OxyR. Cell Chemistry & Biology. 2019;**26**(12):1732-42.e5. DOI: 10.1016/j.chembiol.2019.09.003

[148] Ruegsegger L, Xiao J, Naziripour A, Kanumuambidi T, Brown D, Williams F, et al. Multidrug-resistant gram-negative bacteria in burn patients. Antimicrobial Agents and Chemotherapy. 2022;**66**(9):e0068822. DOI: 10.1128/aac.00688-22

[149] Gong Z, Li J, Luo H, Zhan D, Liu X, Gao C, et al. Low-temperature laminar flow ward for the treatment of multidrug resistance *Acinetobacter baumannii* pneumonia. European Journal of Clinical Microbiology & Infectious Diseases. 2020;**39**(5):877-887. DOI: 10.1007/s10096-019-03790-x

[150] García-Quintanilla M, Caro-Vega JM, Pulido MR, Moreno-Martínez P, Pachón J, McConnell MJ. Inhibition of LpxC increases antibiotic susceptibility in *Acinetobacter baumannii*. Antimicrobial Agents and Chemotherapy. 2016;**60**(8):5076-5079. DOI: 10.1128/aac.00407-16

[151] Lin L, Tan B, Pantapalangkoor P, Ho T, Baquir B, Tomaras A, et al. Inhibition of LpxC protects mice from resistant *Acinetobacter baumannii* by modulating inflammation and enhancing phagocytosis. MBio. 2012;**3**(5):e00312-e00312. DOI: 10.1128/mBio.00312-12

[152] Flamm RK, Rhomberg PR, Jones RN, Farrell DJ. In vitro activity of RX-P873 against *Enterobacteriaceae*, *Pseudomonas aeruginosa*, and *Acinetobacter baumannii*. Antimicrobial Agents and Chemotherapy. 2015;**59**(4):2280-2285. DOI: 10.1128/aac.04840-14

[153] Morrissey I, Magnet S, Genet E, Jeandey P, Marra A, Hawser S, et al. Activity of RX-P873, a novel pyrrolocytosine, against gram-negative bacteria. International Journal of Antimicrobial Agents. 2015;**46**(3):352-354. DOI: 10.1016/j.ijantimicag.2015.06.001

[154] Fitzpatrick MA, Esterly JS, Postelnick MJ, Sutton SH. Successful treatment of extensively drug-resistant *Acinetobacter baumannii* peritoneal dialysis peritonitis with intraperitoneal polymyxin B and ampicillin-sulbactam. The Annals of Pharmacotherapy. 2012;**46**(7-8):e17. DOI: 10.1345/aph.1R086

Chapter 3

Host-Pathogen Interactions in *Acinetobacter baumannii* Infections: Mechanisms of Immune Evasion and Potential Therapeutic Targets

Eunice Damilola Wilkie, Jude Oluwapelumi Alao and Tosin Akin Akinmolayan

Abstract

The book chapter titled "Host–Pathogen Interactions in *Acinetobacter baumannii* Infections: Mechanisms of Immune Evasion and Potential Therapeutic Targets" provides an in-depth exploration of the complex interplay between *A. baumannii*, a notorious multidrug-resistant pathogen, and the host immune system. The chapter will focus on elucidating the mechanisms employed by *A. baumannii* to evade and subvert the immune response, leading to persistent and challenging infections. It will highlight key aspects of the host immune system, including innate and adaptive immunity, pattern-recognition receptors, and immune cell responses, in the context of *A. baumannii* infections. Additionally, the chapter discusses the virulence factors and strategies employed by *A. baumannii* to establish infection, such as biofilm formation and quorum sensing. Importantly, the chapter will explore potential therapeutic targets for combating *A. baumannii* infections, including novel antimicrobial agents, immunotherapies, and host-directed therapies. The comprehensive analysis of host–pathogen interactions and identification of therapeutic strategies presented in this chapter contribute to our understanding of *A. baumannii* infections and pave the way for future research directions and healthcare interventions in combating this formidable pathogen.

Keywords: *A. baumannii*, defense, host–pathogen interactions, immune evasion, therapeutic targets

1. Introduction

1.1 Background and significance of *A. baumannii* infections

A. baumannii has gained notoriety as an emerging nosocomial pathogen, characterized by its rapid development of drug resistance and its affinity for adhering to abiotic surfaces, including medical equipment [1]. This unique ability contributes to the widespread dissemination of the bacterium and presents a formidable challenge in controlling *A. baumannii* infections, particularly ventilator-associated pneumonia in clinical settings [2]. A comprehensive understanding of the intricate host–pathogen interactions during *A. baumannii* infections is needed to combat this elusive pathogen effectively.

The chapter delves into the array of virulence factors employed by *A. baumannii* that are recognized by host innate pattern-recognition receptors. Activation of downstream inflammasomes triggers inflammatory responses, and innate immune effectors are recruited to counter *A. baumannii* infection. This detailed analysis reveals the tug-of-war between the pathogen's virulence factors and the host's immune surveillance, highlighting a complex dance determining the course of the disease.

A. baumannii strategically regulates the expression of various virulence factors to counteract host immune attacks [1, 2]. The chapter illuminates these evasion strategies, providing insights into how the bacterium manipulates the immune landscape. Furthermore, the discussion extends to potential therapeutic targets to combat *A. baumannii* infections. Novel antimicrobial agents, immunotherapies, and host-directed therapies are evaluated for their potential to disrupt the delicate balance of host–pathogen interactions.

The comprehensive analysis presented in this chapter significantly contributes to our understanding of *A. baumannii* infections and paves the way for future research avenues and healthcare interventions. By unraveling the intricate mechanisms of immune evasion and identifying potential therapeutic targets, this chapter empowers the scientific community to combat the challenges posed by *A. baumannii* and devise strategies that promise to improve patient outcomes and address this urgent global health concern.

1.2 Objectives and scope of the book chapter

The primary objective of the book chapter is to provide a comprehensive exploration of the intricate interplay between the multidrug-resistant pathogen *A. baumannii* and the host immune system. The chapter aims to elucidate the mechanisms underlying *A. baumannii*'s ability to evade and subvert the host immune response, thereby establishing persistent and challenging infections. By delving into the complex interactions between the pathogen and the immune system, the chapter contributes to a deeper understanding of *A. baumannii* infections.

The scope of the chapter encompasses various facets of host–pathogen interactions, focusing on innate and adaptive immunity. It covers key components such as pattern-recognition receptors, immune cell responses, and the role of virulence factors in evading immune surveillance. Additionally, the chapter explores the virulence strategies employed by *A. baumannii*, including biofilm formation and quorum sensing. Importantly, the chapter goes beyond elucidating the immune evasion mechanisms and examines potential therapeutic targets for combating *A. baumannii* infections. In addition, it includes a detailed discussion of novel antimicrobial agents, immunotherapies, and host-directed therapies. Overall, the chapter aims to provide a comprehensive analysis that contributes to our understanding of *A. baumannii* infections and paves the way for future research directions and healthcare interventions.

2. Overview of *A. baumannii*

2.1 Taxonomy and classification

A. baumannii belongs to the domain bacteria, phylum *Pseudomonadota*, class *Gammaproteobacteria*, order *Pseudomonadales*, and family *Moraxellaceae* within the genus *Acinetobacter* [1]. The genus includes various species, among which the *A. baumannii* complex is of particular clinical relevance. This complex comprises *A. baumannii*, *Acinetobacter nosocomialis, A. pitii,* and *Acinetobacter calcoaceticus* [2]. Among these, *A. baumannii* is the most clinically significant species within this complex, responsible for various hospital-acquired infections [1].

The *A. calcoaceticus-A. baumannii* complex (ACB complex) is a group of closely related bacterial species within the genus *Acinetobacter*. This complex comprises several species that share genetic similarities and often pose challenges for accurate identification due to their phenotypic similarities [2–4].

A. calcoaceticus (Genomic Species 1) is an environmental species with limited clinical significance. It is part of the ACB complex and is genetically related to other species within the complex. While it is often associated with environmental sources such as soil and water, its role in clinical infections is not as well-defined [3, 4].

A. baumannii (Genomic Species 2) is the most clinically important species within the ACB complex. It is a gram-negative bacterium responsible for various infections, particularly in healthcare settings. *A. baumannii* is associated with multidrug resistance, making it challenging to treat. It is a major cause of nosocomial outbreaks and has been extensively studied due to its impact on patient health [1, 2].

Acinetobacter pittii (Genomic Species 3) is another member of the ACB complex and is closely related to *A. baumannii.* It shares genetic similarities with other species in the complex, making accurate identification difficult using conventional methods. It has been isolated from clinical specimens and is associated with healthcare-associated infections [3, 4].

A. nosocomialis (Genomic Species 13TU) is part of the ACB complex and is closely related to other species within the complex. It shares genetic traits with *A. baumannii* and *A. pittii*, making it challenging to distinguish phenotypically. Like other species in the complex, *A. nosocomialis* has been isolated from clinical specimens and is associated with nosocomial infections [2, 3].

Although not originally included in the ACB complex, *A. seifertii* has been proposed for inclusion within the complex. It was previously called Acinetobacter genomic species "close to 13TU." This species has been isolated from clinical specimens and contributes to the complexity of *Acinetobacter* species identification [3, 4].

Acinetobacter lactucae (Synonym of A. dijkshoorniae) formerly known as Acinetobacter NB14, is closely related to *A. pittii* and *A. nosocomialis.* It has been identified as a high-priority pathogen, especially in intensive care units. *A. lactucae* is associated with clinical infections, and its inclusion within the ACB complex adds to the challenges of accurate identification [3, 4].

A. baumannii is a short, almost round, rod-shaped (coccobacillus) Gram-negative bacterium. It lacks flagella for locomotion but exhibits twitching or swarming motility, possibly due to type IV pili or exopolysaccharide activity. While other species of the *Acinetobacter* genus are often found in soil [1, 2], *A. baumannii* is primarily isolated from hospital environments [1], making it an important nosocomial pathogen. It can be an opportunistic pathogen, particularly affecting individuals with compromised immune systems [1–3].

The taxonomy of the *Acinetobacter* genus has evolved, leading to the recognition of distinct species within the *A. calcoaceticus-A. baumannii* complex [3]. Initial taxonomic studies in the mid-1980s identified *A. baumannii* as a novel species, separate from other Acinetobacter species [3, 4]. Further refinements in taxonomy included the proposal of new species, such as *A. pittii* "Genomic Species 3" and *A. nosocomialis*

"Genomic Species 13TU" [1, 4]. The bacterium's ability to adapt and thrive in diverse environments underscores its clinical significance and challenges in infection control.

2.2 Clinical relevance and epidemiology

A. baumannii is a prominent pathogenic bacterium associated with various healthcare-associated infections, posing significant challenges to medical communities worldwide [1, 2]. Its clinical significance is underscored by its ability to cause a wide range of infections, its propensity for antibiotic resistance, and its capacity for persistence in hospital environments [5]. It commonly colonizes respiratory secretions, wounds, urine, and various medical equipment within hospital environments [5]. The risk of acquiring an *A. baumannii* infection is heightened in individuals with prior antibiotic exposure, intensive care unit (ICU) admissions, central venous catheter usage, and mechanical ventilation or hemodialysis [6]. Infections often target organ systems with high fluid content, such as the respiratory tract, cerebrospinal fluid, peritoneal fluid, and urinary tract [3]. Notably, outbreaks of *Acinetobacter* infections, particularly pneumonia, have been reported in healthcare settings [1, 7, 8].

The epidemiology of *A. baumannii* presents a significant public health concern, particularly within healthcare settings. This ubiquitous pathogen is capable of causing both community and healthcare-associated infections (HAIs), with the latter being the more common form [9]. *A. baumannii* has gained attention due to its extensive antimicrobial resistance and ability to initiate large, often multi-facility nosocomial outbreaks [5]. These outbreaks are facilitated by its tolerance to desiccation and its multidrug resistance, allowing it to persist in hospital environments [10].

The epidemiology of *A. baumannii* infections is complex, with the coexistence of both epidemic and endemic diseases. Epidemic infections can lead to outbreaks, while endemic infections are often fueled by the selective pressure of antimicrobials [11]. Notably, severe *A. baumannii* infections, such as bacteremia or pneumonia in intensive care unit patients undergoing intubation, are not associated with higher attributable mortality rates or increased hospital stays [12]. The pathogen mainly causes pulmonary, urinary tract, bloodstream, or surgical wound infections, with invasive procedures and broad-spectrum antimicrobial use being significant risk factors [4, 5]. Despite its clinical importance, knowledge about *A. baumannii* is less developed than other pathogens, and accurate identification remains challenging [5]. Nevertheless, the organism's ability to accumulate antimicrobial resistance mechanisms, resistance to desiccation, and propensity to cause outbreaks make it a noteworthy and challenging pathogen in healthcare settings.

2.3 Virulence factors and pathogenicity

Virulence factors are crucial determinants that contribute to *A. baumannii*'s ability to establish infections, evade host defenses, and cause disease. Several virulence factors have been identified, shedding light on this bacterium's pathogenesis and virulence mechanisms.

One notable virulence factor is the presence of efflux pumps, which contribute to antibiotic resistance and facilitate the extrusion of antibiotics, limiting their effectiveness [13]. Additionally, β-lactamases and aminoglycoside-modifying enzymes significantly confer antibiotic resistance, further enhancing the bacterium's ability to survive and cause infections [14].

Biofilm formation is another important virulence factor that enables *A. baumannii* to adhere to surfaces, including medical devices and equipment, contributing to its persistence in healthcare environments [10, 15]. The Bap protein and the csu locus are associated with biofilm production and pathogenicity, allowing *A. baumannii* to colonize and establish infections on medical surfaces [10].

Furthermore, iron acquisition systems are critical virulence factors that facilitate the acquisition of iron, an essential nutrient for bacterial growth, from the host environment [16]. Iron is crucial for bacterial survival and proliferation, and *A. baumannii* has developed mechanisms to scavenge iron from the host to support its growth during infection [17].

The outer membrane protein OmpA, phospholipases, membrane polysaccharide components, penicillin-binding proteins, and outer membrane vesicles are additional virulence factors identified in *A. baumannii* and contribute to its pathogenesis [18]. These factors play roles in host immune responses, bacterial adherence, and evasion of host defenses.

The aforementioned virulence factors confer *A. baumannii* as a formidable opportunistic pathogen known for its ability to cause severe nosocomial infections, particularly in intensive care units (ICUs) [19–21]. Its emergence as a public health threat is underscored by its escalating antibiotic resistance and its ability to cause various clinical manifestations, including pneumonia, septicemia, and meningitis [3].

A critical facet of *A. baumannii*'s pathogenicity is its ability to trigger a robust immune response upon infection. Toll-like Receptor 4 (TLR4) serves as a key pathogen recognition receptor, inducing the production of inflammatory cytokines, such as IL-6 and TNF-α, upon *A. baumannii* infection [22]. Activating the inflammasome pathway leads to pyroptosis and the release of pro-inflammatory cytokines, contributing to the host's defense mechanisms against the infection [23].

A recent study has identified that *A. baumannii* secretes a bioactive lipid that triggers inflammatory signaling and cell death [22], further highlighting its capacity to induce immune responses. Specific virulence factors, such as phospholipases and outer membrane proteins, also contribute to its ability to adhere to host cells and evade immune recognition [24].

The epidemiology of *A. baumannii* infections suggests that it can cause outbreaks in healthcare settings, particularly ICUs. Cross-contamination between patients and the environment plays a significant role in its transmission [3]. This fact emphasizes the importance of infection control measures to prevent its spread and reduce hospital-acquired infections [3].

The pathogenicity of *A. baumannii* is multifaceted, encompassing antibiotic resistance, immune activation, biofilm formation, and virulence factor secretion. Its ability to trigger immune responses while evading host defenses contributes to its clinical impact and challenges in treatment.

3. Host immune responses to *A. baumannii*

3.1 Innate immune responses

3.1.1 Recognition and activation of innate immune cells

The host's innate immune responses are pivotal in the initial recognition and defense against *A. baumannii* infection. Innate immune responses are orchestrated by a complex interplay between pattern-recognition receptors (PRRs) and various immune effectors [25]. Understanding these interactions is critical for developing novel therapeutic strategies, including vaccines and immunotherapeutics, to combat *A. baumannii* infections.

Recognition of *A. baumannii* by PRRs, such as Toll-like receptors (TLRs), initiates a cascade of events leading to the production of inflammatory cytokines and chemokines [26]. These signaling molecules recruit innate immune effectors, including neutrophils and macrophages, to the site of infection. Neutrophils, in particular, play a crucial role in the control of *A. baumannii* infections [27]. They are rapidly recruited to the site of infection and contribute to bacterial clearance through phagocytosis and the release of antimicrobial peptides and reactive oxygen species.

However, *A. baumannii* has evolved mechanisms to evade immune responses and establish infections. Its ability to develop antibiotic resistance further complicates treatment strategies, highlighting the need for alternative approaches such as immunomodulation [28].

3.1.2 Inflammatory cytokine production

The host's innate immune responses play a critical role in recognizing and responding to *A. baumannii* infection, producing inflammatory cytokines and chemokines that orchestrate the immune defense against the pathogen [26].

The host's PRRs recognize pathogen-associated molecular patterns (PAMPs) on the bacterium's surface. This recognition triggers a cascade of events that lead to the activation of downstream signaling pathways, ultimately producing pro-inflammatory cytokines and chemokines [27].

Inflammatory cytokines, such as interleukin-1 (IL-1), tumor necrosis factor-alpha (TNF-α), and interleukin-6 (IL-6), are key mediators of the immune response against *A. baumannii* infection [28]. These cytokines play pivotal roles in promoting inflammation, recruiting immune cells to the site of infection, and enhancing immune cell activation. For instance, neutrophils, essential players in controlling *A. baumannii* infection, are rapidly recruited to the site of infection in response to cytokine signals [29]. Neutrophils contribute to bacterial clearance through phagocytosis and the release of antimicrobial peptides and reactive oxygen species [30].

However, *A. baumannii* has evolved mechanisms to evade immune responses, including modulating the expression of virulence factors to counteract host immune attacks [31]. This tug-of-war between the bacterium and the host's immune system underscores the complexity of the immune response against *A. baumannii* infection.

3.1.3 Phagocytosis and intracellular killing

Phagocytosis, the process by which immune cells engulf and internalize pathogens, is pivotal in the initial defense against *A. baumannii*. Neutrophils, macrophages, and other professional phagocytes are essential effectors in host defense against this bacterium [26]. Neutrophils, in particular, are rapid responders recruited to the site of infection and are crucial for controlling *A. baumannii* infections [26]. Upon encountering *A. baumannii*, neutrophils undergo activation, leading to flattening and the extension of pseudopods, which initiate phagocytosis [32]. This process involves recognizing bacterial components through pattern-recognition receptors, such as TLRs, on the surface of neutrophils. These interactions trigger bactericidal mechanisms, including oxidative bursts and the production of cytokines and chemokines, amplifying the immune response against the pathogen [33]. Furthermore, neutrophils have been shown to release neutrophil extracellular traps (NETs), web-like structures composed of DNA, histones, and antimicrobial proteins, as part of their defense against *A. baumannii* [32].

In addition to phagocytosis, intracellular killing mechanisms are critical for neutralizing *A. baumannii* within immune cells. Once phagocytosed, immune cells destroy the engulfed bacteria through various means. Professional phagocytes can generate reactive oxygen species (ROS) through oxidative burst, which is toxic to internalized pathogens [33]. These ROS contribute to the bactericidal activity of immune cells, aiding in the elimination of *A. baumannii* [33]. Moreover, the production of antimicrobial peptides and enzymes within phagolysosomes further enhances the intracellular killing of *A. baumannii* [33].

While neutrophils and other immune cells play a crucial role in phagocytosis and intracellular killing, the ability of *A. baumannii* to survive within host cells and manipulate immune responses poses challenges in combating infections caused by this pathogen.

3.2 Adaptive immune responses

3.2.1 T cell-mediated responses

T cell-mediated responses, including CD4+ helper T cells and CD8+ cytotoxic T cells, are essential components of the adaptive immune system's defense against *A. baumannii*. These T cells recognize specific antigens presented by antigen-presenting cells (APCs) and respond by proliferating and differentiating into armed effector T cells. CD4+ T cells help other immune cells, such as B cells and macrophages, enhance the immune response. CD8+ T cells target and eliminate *A. baumannii*-infected host cells [26].

Despite the importance of T cell-mediated responses, understanding the host immune interaction with *A. baumannii* still needs to be completed. Developing effective vaccines and immunotherapies to combat *A. baumannii* infections requires a deeper comprehension of the host immune mechanisms, identifying key virulence factors targeted by the immune system, and the modulation of T cell responses to enhance their efficacy against this pathogen. The ongoing efforts to elucidate the immune response to *A. baumannii* will contribute to developing innovative strategies to mitigate its impact on global health.

3.2.2 B cell-mediated responses

The role of B cell-mediated responses in combating *A. baumannii* infections is paramount. B cells, a crucial component of the adaptive immune system, contribute to the defense against pathogens by producing antibodies and participating in immune memory.

B cells play a central role in recognizing specific antigens presented by *A. baumannii* [33]. Upon encountering these antigens, B cells become activated and undergo clonal expansion, producing antibodies specifically tailored to bind to the pathogen. These antibodies can neutralize *A. baumannii* by preventing its interaction with host cells and opsonizing the bacterium for phagocytosis by innate immune cells [34].

The antibodies produced by B cells can initiate various effector mechanisms that contribute to the clearance of *A. baumannii* infections. These mechanisms include complement activation, which enhances opsonization and lysis of the pathogen, and antibody-dependent cellular cytotoxicity (ADCC), where immune cells such as neutrophils and macrophages recognize and eliminate antibody-bound *A. baumannii* [34].

One of the key functions of B cells is to establish immunological memory. Memory B cells are long-lived and can rapidly respond to re-infection with *A. baumannii*. Upon re-exposure to the pathogen, memory B cells can quickly differentiate into antibody-secreting plasma cells, leading to a faster and more robust immune response. This memory response is essential for preventing recurrent infections and providing long-term protection [34].

Despite the critical role of B cell-mediated responses, challenges remain in fully understanding their specific interactions with *A. baumannii* antigens and elucidating the antigenic targets that elicit protective B cell responses and characterizing the antibody repertoire generated during *A. baumannii* infection will contribute to the development of effective vaccines and immunotherapies. Additionally, the impact of *A. baumannii*'s ability to adapt and regulate virulence factor expression on B cell responses requires further investigation.

3.2.3 Antibody production and opsonization

Antibodies, also known as immunoglobulins (Ig), are produced by B lymphocytes in response to the presence of antigens, such as *A. baumannii* components. Upon exposure to *A. baumannii*, B cells recognize specific antigens, leading to their activation and subsequent differentiation into plasma cells. These plasma cells secrete antibodies tailored to target *A. baumannii* antigens, facilitating their neutralization and removal from the body [35, 36]. The production of antibodies, particularly IgG, is a key feature of the adaptive immune response against *A. baumannii* infections.

Opsonization is a crucial process by which antibodies bind to pathogens, marking them for recognition and engulfment by immune cells such as phagocytes. Antibodies attached to *A. baumannii* enhance the efficiency of phagocytosis by facilitating the interaction between the pathogen and immune cells. This opsonic effect improves the clearance of *A. baumannii* from the host's bloodstream and infected tissues. Opsonization is particularly important in countering *A. baumannii*'s ability to evade the host immune response through mechanisms such as capsule formation and outer membrane protein variation [34, 35].

Immunization strategies involving *A. baumannii* antigens, such as outer membrane vesicles (OMVs) or capsular polysaccharides, have shown promise in inducing robust antibody responses [35]. Immunization with *A. baumannii* OMVs has been demonstrated to elicit high levels of IgG antibodies, which are associated with opsonization and improved antibiotic sensitivity of the pathogen [35]. These antibodies can enhance the susceptibility of *A. baumannii* to antibiotics, potentially enhancing the effectiveness of antibiotic treatments [35].

4. Mechanisms of immune evasion by *A. baumannii*

4.1 Capsule and outer membrane proteins

The capsule of *A. baumannii* is a protective polysaccharide layer that envelops the bacterium, enabling it to evade recognition by host immune cells. This structure hampers opsonization, a process in which antibodies or complement proteins coat the pathogen, marking it for phagocytosis by immune cells. By masking surface antigens and inhibiting complement deposition, the capsule shields *A. baumannii* from immune detection and subsequent destruction [37].

A. baumannii employs outer membrane proteins (OMPs) as versatile tools to modulate interactions with the host immune system. These OMPs are pivotal in mediating adhesion, invasion, and immune evasion. Through antigenic variation and phase variation, *A. baumannii* can alter the expression of specific OMPs, evading immune surveillance and memory. Additionally, some OMPs have been shown to interact with host receptors, thereby dampening immune responses and promoting bacterial survival [31].

4.2 Efflux pumps and antibiotic resistance

Efflux pumps are integral membrane proteins that transport many molecules, including antibiotics, out of bacterial cells. *A. baumannii* employs efflux pumps to expel antibiotics from within the bacterial cell, thereby reducing intracellular drug concentrations and rendering antibiotics less effective. Efflux pumps contribute to multidrug resistance (MDR) in *A. baumannii*, enabling the bacterium to survive exposure to various antibiotics, including aminoglycosides, fluoroquinolones, and beta-lactams [13, 27].

Several classes of efflux pumps are associated with *A. baumannii*'s antibiotic resistance. Notably, the major facilitator superfamily (MFS), resistance-nodulation cell division (RND) family, small multidrug resistance (SMR) family, and multidrug and toxic compound extrusion (MATE) family of efflux pumps are implicated in the bacterium's ability to expel antibiotics and evade host immune responses [13, 27]. These pumps have three main components: the outer membrane channel, the periplasmic lipoprotein, and the inner membrane transporter.

Efflux pumps reduce antibiotic susceptibility by preventing antibiotics from accumulating within *A. baumannii* cells. This phenomenon leads to elevated minimum inhibitory concentrations (MICs) of antibiotics required to inhibit bacterial growth. Consequently, the bacterium becomes more resistant to antibiotic treatments, limiting the effectiveness of conventional therapeutic approaches [13, 27].

4.3 Biofilm formation

Biofilms are complex communities of bacterial cells encased within a self-produced extracellular matrix. This matrix, primarily composed of polysaccharides, proteins, and DNA, protects bacteria from external threats, including host immune cells and antibiotics. *A. baumannii*'s ability to form biofilms allows it to attach to biotic and abiotic surfaces, making medical devices and equipment potential reservoirs for infection [10, 15, 38].

Biofilm formation enables *A. baumannii* to evade the host immune response through multiple mechanisms. The biofilm matrix acts as a physical barrier that hinders the penetration of immune cells and antibodies, thereby reducing the efficacy of the immune system's defense mechanisms. Additionally, the altered physiology of bacterial cells within the biofilm contributes to decreased susceptibility to immune clearance. Immune cells, such as neutrophils and macrophages, struggle to effectively target and eliminate bacteria embedded within the biofilm structure [10, 15, 38].

A. baumannii biofilms are frequently associated with chronic infections, particularly those involving medical devices like catheters and ventilators. These infections are challenging to treat due to the inherent resistance of biofilm-embedded bacteria to antibiotics. The biofilm matrix provides a protective environment that shields bacteria from the effects of antibiotics and prevents their effective eradication. As a result, chronic infections caused by *A. baumannii* biofilms can persist despite antibiotic treatment, leading to prolonged patient suffering and increased healthcare costs [15, 38].

4.4 Modification of surface structures and antigenic variation

A. baumannii employs various strategies to modify its surface structures, effectively masking its presence from the host immune system. One of the key modifications is the alteration of lipopolysaccharides (LPS) and OMPs, which are major targets for host immune recognition. By modifying these surface molecules, *A. baumannii* can evade detection by immune cells and antibodies, reducing the effectiveness of the immune response. Additionally, *A. baumannii* may shed OMVs containing modified surface components, further contributing to immune evasion [31, 39].

Antigenic variation is a sophisticated strategy employed by *A. baumannii* to continually alter its surface antigens, making it difficult for the host immune system to recognize and mount an effective response. *A. baumannii* possesses a diverse repertoire of surface antigens, such as pili and fimbriae, which can undergo rapid changes through genetic recombination and mutation. This dynamic antigenic variation hinders the host's ability to generate a robust and lasting immune response, allowing *A. baumannii* to evade immune surveillance and persist within the host [31, 37, 39].

Modifying surface structures and antigenic variation collectively contribute to *A. baumannii*'s ability to escape immune recognition and clearance. These mechanisms limit the host's ability to develop a strong and sustained immune response against the bacterium. As a result, *A. baumannii* can persist within the host, leading to chronic infections that are challenging to treat with conventional antibiotics. This persistence is particularly problematic in healthcare settings, where *A. baumannii* can cause ventilator-associated pneumonia and other hospital-acquired infections [31, 37, 39].

4.5 Suppression of immune signaling pathways

A. baumannii utilizes several mechanisms to dampen host immune signaling pathways, impairing the immune response and promoting its survival within the host environment. One key strategy involves interference with PRRs, crucial in initiating immune responses upon pathogen detection. By inhibiting PRR signaling, *A. baumannii* can thwart the activation of immune cascades, reducing the recruitment of immune effectors and impeding the production of inflammatory cytokines and chemokines necessary for an effective immune response [24, 26].

Neutrophils are essential components of the innate immune system and play a critical role in combatting bacterial infections. *A. baumannii* employs strategies to counteract neutrophil responses, impairing their recruitment and effector functions. Studies have shown that *A. baumannii* can interfere with neutrophil recruitment to the site of infection, leading to delayed reactions and reduced bactericidal activity. Furthermore, the bacterium can modulate cytokine and chemokine production, hindering the optimal activation of neutrophils and other immune cells required for effective bacterial clearance [24, 26, 33].

A. baumannii employs a multifaceted approach to evade host immune responses, including suppressing key signaling pathways involved in immune activation. This evasion strategy impairs the initial recognition of the pathogen by the host and dampens the subsequent immune cascade required for efficient bacterial clearance.

By targeting these immune signaling pathways, *A. baumannii* can establish chronic infections and evade host defenses, contributing to its persistence and clinical significance as a nosocomial pathogen [24, 26].

5. Impact of host: pathogen interactions on disease outcome

5.1 Factors influencing disease severity and prognosis

The severity of *A. baumannii* infections has a direct impact on patient prognosis. Studies have shown that higher Acute Physiology and Chronic Health Evaluation (APACHE) II scores indicate that patients with more severe disorders are at increased risk of mortality. Patients with underlying severe comorbidities, such as hematologic malignancies, are particularly vulnerable to poor outcomes in the presence of *A. baumannii* infections [40–42].

Appropriate antimicrobial therapy is a critical determinant of patient outcomes in *A. baumannii* infections. Studies have demonstrated that timely and effective antimicrobial treatment reduces mortality rates, particularly in severely ill patients. However, the impact of antimicrobial therapy may vary based on the severity of infection, underlying conditions, and other risk factors [40, 41].

Several factors contribute to the severity and prognosis of *A. baumannii* infections. These include the presence of neutropenia, which weakens the immune response, and the use of invasive procedures, which can introduce and spread diseases. Additionally, the prior use of specific antibiotics, such as carbapenems, has been identified as a risk factor for poor outcomes. Mechanical ventilation and initial immunosuppression are also associated with increased mortality rates in *A. baumannii* bloodstream infections [41].

MDR is a significant concern in *A. baumannii* infections, potentially limiting treatment options and contributing to poorer outcomes. While MDR may not always be a direct risk factor for mortality, it can impact the choice of appropriate antimicrobial therapy, potentially leading to treatment failure and increased mortality rates [42].

5.2 Host genetic susceptibility

Host genetic susceptibility refers to inherited gene variations that can affect an individual's susceptibility to infections and their ability to mount an effective immune response. The genetic diversity among individuals can impact the interaction between *A. baumannii* and the host's immune system. Some genetic variations may enhance the host's ability to recognize and combat the pathogen, while others may compromise the immune response and increase susceptibility to infection.

Host genetic factors play a role in shaping both innate and adaptive immune responses to *A. baumannii* infections. Variations in genes encoding immune receptors, cytokines, and other immune-related molecules can affect the intensity and effectiveness of the immune response. For example, genetic variations in TLRs or cytokines may influence the recognition of *A. baumannii* and subsequent activation of immune signaling pathways. These genetic differences can impact the production of pro-inflammatory cytokines, chemokines, and other immune mediators, which affect the recruitment and activation of immune cells [39, 43].

A. baumannii employs various virulence factors to establish infection and evade host immune responses. The host's genetic background can influence these virulence factors' effectiveness. For instance, host cell surface receptors or signaling molecule variations may affect the pathogen's ability to adhere to and invade host cells. Additionally, genetic variations in the host may impact the immune system's recognition of specific virulence factors, influencing the overall immune response to the infection [39, 43].

6. Potential therapeutic targets for *A. baumannii* infections

6.1 Antibiotic resistance mechanisms and novel antimicrobial strategies

The increasing prevalence of MDR, extensive drug-resistant (XDR), and even pan-drug-resistant (PDR) strains of *A. baumannii* has raised concerns about limited treatment options and the need for novel antimicrobial strategies [44, 45].

The development of antibiotic resistance in *A. baumannii* is a complex process driven by various genetic and physiological factors. The resistance mechanisms primarily involve regulating antibiotic transportation through bacterial membranes, alteration of the antibiotic target site, and enzymatic modifications that neutralize antibiotics.

The rise of MDR, XDR, and PDR *A. baumannii* strains can be attributed to extensive antibiotic abuse and poor stewardship in healthcare settings. Long hospitalization stays, catheters, mechanical ventilation, and compromised immune systems further contribute to the emergence of resistant strains [44]. In recent years, there has been a growing awareness of the need for appropriate antibiotic use, infection prevention, and surveillance strategies to curb the spread of antibiotic resistance.

Advances in next-generation sequencing techniques have revolutionized the diagnosis of severe *A. baumannii* infections. These techniques allow for the rapid identification of specific resistance genes, enabling timely diagnosis and the design of personalized therapeutic regimens based on the pathogen's resistance profile [44]. Tailoring treatment to the identified resistance mechanisms enhances the likelihood of successful outcomes and reduces the risk of treatment failure.

Researchers are exploring novel antimicrobial strategies to combat *A. baumannii* infections, especially those caused by MDR and XDR strains. One such approach involves the development of alternative antibiotics or antimicrobial agents that target different bacterial pathways, reducing the likelihood of cross-resistance [45]. Efforts are also underway to investigate combination therapies that synergistically enhance the efficacy of existing antibiotics, potentially overcoming resistance mechanisms.

6.2 Targeting virulence factors and host: pathogen interactions

A. baumannii deploys a variety of virulence factors to establish infections and evade host defenses. Targeting these virulence mechanisms presents a potential strategy to disrupt the pathogenicity of *A. baumannii*.

Understanding the interactions between *A. baumannii* and the host immune system is essential for developing effective therapeutic interventions. *A. baumannii* has evolved mechanisms to evade host immune responses and establish persistent infections. By disrupting these interactions, researchers aim to enhance the host's ability to clear the infection and improve treatment outcomes [45, 46].

Researchers are investigating various approaches to target virulence factors and host–pathogen interactions in *A. baumannii* infections. These strategies include the development of new antimicrobial agents that inhibit essential bacterial functions and therapies that specifically disrupt virulence mechanisms without killing the bacterium. For example, inhibiting OmpA, a key virulence factor, could weaken the bacterium's ability to form biofilms and evade immune responses, making it more susceptible to clearance by the host [46].

6.3 Immunotherapeutic approaches

The emergence of MDR and XDR *A. baumannii* strains has led to limited treatment options, with traditional antibiotics becoming increasingly ineffective. This resistance is attributed to the mechanisms mentioned earlier. Consequently, alternative strategies, including immunotherapeutic approaches, are being explored to address the growing threat of *A. baumannii* infections.

Immunization trials are being considered as a promising avenue for combatting *A. baumannii* infections. Researchers are focusing on developing vaccines that target specific antigens or epitopes associated with the pathogen. Several antigens and peptides have been proposed for active and passive immunizations [47].

Monoclonal antibody (MAb) therapy has emerged as a promising immunotherapeutic strategy against *A. baumannii* infections. MAbs are designed to recognize and neutralize bacterial targets specifically. Researchers have developed MAbs that target different components of *A. baumannii*, such as the bacterial capsule, to enhance opsonophagocytosis and clearance by immune cells [48]. These MAbs have demonstrated efficacy in murine models, significantly improving survival rates and reducing bacterial loads [48]. Furthermore, combining monoclonal antibody therapy with traditional antibiotics, such as colistin, has shown synergistic effects and improved protection [48].

While immunotherapeutic approaches, including monoclonal antibody therapy and vaccine development, hold promise for addressing *A. baumannii* infections, challenges remain. Designing vaccines that provide broad protection against various strains and do not affect the host microbiota or proteome is complex [47]. Additionally, the clinical translation of immunotherapeutic strategies requires rigorous preclinical testing and validation.

7. Conclusion

This book chapter has provided a comprehensive overview of host–pathogen interactions in *A. baumannii* infections, uncovering immune evasion mechanisms and potential therapeutic targets. While challenges persist, the remarkable progress in understanding these interactions offers hope for innovative treatments and strategies to combat *A. baumannii* infections and improve patient outcomes.

As we conclude this exploration, several implications for future research and therapeutic interventions emerge. Firstly, the need for a deeper understanding of the host immune response and the molecular mechanisms of *A. baumannii*'s immune evasion strategies remains paramount. Identifying specific virulence factors and the regulatory networks that govern their expression could provide new targets for intervention.

Advancing research should also focus on unraveling the dynamics of multidrug resistance in *A. baumannii* and developing innovative strategies to circumvent resistance mechanisms. The application of cutting-edge technologies, such as genomics and proteomics, holds promise in identifying novel therapeutic targets and potential biomarkers for early detection and prognosis.

The potential of immunomodulatory agents and host-directed therapies in therapeutic interventions should be rigorously explored. Designing targeted strategies that enhance the host's immune response while inhibiting *A. baumannii*'s immune evasion mechanisms could revolutionize treatment outcomes.

Collaboration between interdisciplinary teams, including microbiologists, immunologists, pharmacologists, and clinicians, will be instrumental in translating research findings into effective therapeutic interventions. Integrating computational modeling and artificial intelligence could expedite drug discovery and enhance our understanding of complex host–pathogen interactions.

Conflict of interest

The authors declare no conflict of interest.

Author details

Eunice Damilola Wilkie*, Jude Oluwapelumi Alao and Tosin Akin Akinmolayan
Department of Microbiology, Adeleke University, Nigeria

*Address all correspondence to: wilkie.eunice@adelekeuniversity.edu.ng

References

[1] Almasaudi SB. *Acinetobacter spp*. as nosocomial pathogens: Epidemiology and resistance features. Saudi Journal of Biological Sciences. 2018;**25**(3):586-596. DOI: 10.1016/j.sjbs.2016.02.009

[2] Vijayakumar S, Biswas I, Veeraraghavan B. Accurate identification of clinically important *Acinetobacter spp.*: An update. Future Science OA. 2019;**5**(6):FSO395. DOI: 10.2144/fsoa-2018-0127

[3] Peleg AY, Seifert H, Paterson DL. *Acinetobacter baumannii*: Emergence of a successful pathogen. Clinical Microbiology Reviews. 2008;**21**(3):538-582. DOI: 10.1128/CMR.00058-07

[4] Ayoub Moubareck C, Hammoudi HD. Insights into *Acinetobacter baumannii*: A review of microbiological, virulence, and resistance traits in a threatening nosocomial pathogen. Antibiotics (Basel). 2020;**9**(3):119. DOI: 10.3390/antibiotics9030119

[5] Nocera FP, Attili AR, De Martino L. *Acinetobacter baumannii*: Its clinical significance in human and veterinary medicine. Pathogens. 2021;**10**(2):127. DOI: 10.3390/pathogens10020127

[6] Aboshakwa AM, Lalla U, Irusen EM, Koegelenberg CFN. *Acinetobacter baumannii* infection in a medical intensive care unit: The impact of strict infection control. African Journal of Thoracic and Critical Care Medicine Critical Care Medicine. 2019;**25**(1). DOI: 10.7196/AJTCCM.2019.v25i1.239

[7] Hartzell JD, Kim AS, Kortepeter MG, Moran KA. *Acinetobacter* pneumonia: A review. MedGenMed. 2007;**9**(3):4

[8] Agyepong N, Fordjour F, Owusu-Ofori A. Multidrug-resistant *Acinetobacter baumannii* in healthcare settings in Africa. Frontiers in Tropical Diseases. 2023;**4**:1110125

[9] Fournier PE, Richet H. The epidemiology and control of *Acinetobacter baumannii* in health care facilities. Clinical Infectious Diseases. 2006;**42**(5):692-699. DOI: 10.1086/500202 Epub 2006 Jan 26

[10] Gedefie A, Demsis W, Ashagrie M, Kassa Y, Tesfaye M, Tilahun M, et al. *Acinetobacter baumannii* biofilm formation and its role in disease pathogenesis: A review. Infection and Drug Resistance. 2021;**14**:3711-3719. DOI: 10.2147/IDR.S332051

[11] Fournier PE, Richet H, Weinstein RA. The epidemiology and control of *Acinetobacter baumannii* in health care facilities. Clinical Infectious Diseases. 2006;**42**(5):692-699

[12] Alrahmany D, Omar AF, Alreesi A, Harb G, Ghazi IM. *Acinetobacter baumannii* infection-related mortality in hospitalized patients: Risk factors and potential targets for clinical and antimicrobial stewardship interventions. Antibiotics (Basel). 2022;**11**(8):1086. DOI: 10.3390/antibiotics11081086

[13] Abdi SN, Ghotaslou R, Ganbarov K, Mobed A, Tanomand A, Yousefi M, et al. *Acinetobacter baumannii* efflux pumps and antibiotic resistance. Infection and Drug Resistance. 2020;**13**:423-434. DOI: 10.2147/IDR.S228089

[14] Lin MF, Lan CY. Antimicrobial resistance in *Acinetobacter baumannii*: From bench to bedside. World Journal of Clinical Cases. 2014;**2**(12):787-814. DOI: 10.12998/wjcc.v2.i12.787

[15] Upmanyu K, Haq QMR, Singh R. Factors mediating *Acinetobacter baumannii* biofilm formation: Opportunities for developing therapeutics. Current Research in Microbial Sciences. 2022;**3**:100131. DOI: 10.1016/j.crmicr.2022.100131

[16] Caza M, Kronstad JW. Shared and distinct mechanisms of iron acquisition by bacterial and fungal pathogens of humans. Frontiers in Cellular and Infection Microbiology. 2013;**3**:80. DOI: 10.3389/fcimb.2013.00080

[17] Runci F, Gentile V, Frangipani E, Rampioni G, Leoni L, Lucidi M, et al. Contribution of active iron uptake to *Acinetobacter baumannii* pathogenicity. Infection and Immunity. 2019;**87**(4): e00755-e00718. DOI: 10.1128/IAI.00755-18

[18] Uppalapati SR, Sett A, Pathania R. The outer membrane proteins OmpA, CarO, and OprD of *Acinetobacter baumannii* confer a two-pronged defense in facilitating its success as a potent human pathogen. Frontiers in Microbiology. 2020;**11**:589234

[19] Camp C, Tatum OL. A review of *Acinetobacter baumannii* as a highly successful pathogen in times of war. Laboratoriums Medizin. 2010;**41**(11):649-657

[20] Howard A, O'Donoghue M, Feeney A, Sleator RD. *Acinetobacter baumannii*: An emerging opportunistic pathogen. Virulence. 2012;**3**(3):243-250. DOI: 10.4161/viru.19700 Epub 2012 May 1

[21] Ma C, McClean S. Mapping global prevalence of *Acinetobacter baumannii* and recent vaccine development to tackle it. Vaccines (Basel). 2021;**9**(6):570. DOI: 10.3390/vaccines9060570

[22] Tiku V, Kew C, Kofoed EM, Peng Y, Dikic I, Tan MW. *Acinetobacter baumannii* secretes a bioactive lipid that triggers inflammatory Signaling and cell death. Frontiers in Microbiology. 2022;**13**:870101. DOI: 10.3389/fmicb.2022.870101

[23] Man SM, Karki R, Kanneganti TD. Molecular mechanisms and functions of pyroptosis, inflammatory caspases and inflammasomes in infectious diseases. Immunological Reviews. 2017;**277**(1):61-75. DOI: 10.1111/imr.12534

[24] Sarshar M, Behzadi P, Scribano D, Palamara AT, Ambrosi C. *Acinetobacter baumannii*: An ancient commensal with weapons of a pathogen. Pathogens. 2021;**10**(4):387. DOI: 10.3390/pathogens10040387

[25] Aristizábal B, González Á. Innate immune system. In: Anaya JM, Shoenfeld Y, Rojas-Villarraga A, et al., editors. Autoimmunity: From Bench to Bedside [Internet]. Bogota (Colombia): El Rosario University Press; 2013 Chapter 2. Available from: https://www.ncbi.nlm.nih.gov/books/NBK459455/

[26] Chen W. Host innate immune responses to *Acinetobacter baumannii* infection. Frontiers in Cellular and Infection Microbiology. 2020;**10**:486. DOI: 10.3389/fcimb.2020.00486

[27] Lee CR, Lee JH, Park M, Park KS, Bae IK, Kim YB, et al. Biology of *Acinetobacter baumannii*: Pathogenesis, antibiotic resistance mechanisms, and prospective treatment options. Frontiers in Cellular and Infection Microbiology. 2017;**7**. [Internet] Available from: http://journal.frontiersin.org/article/10.3389/fcimb.2017.00055/full

[28] Korkmaz FT, Traber KE. Innate immune responses in pneumonia. Pneumonia. 2023;**15**(1):4

[29] Liu Z, Xu W. Neutrophil and macrophage response in *Acinetobacter*

Baumannii infection and their relationship to lung injury. Frontiers in Cellular and Infection Microbiology. 2022;**12**:890511. DOI: 10.3389/fcimb.2022.890511

[30] Teng TS, Ji AL, Ji XY, Li YZ. Neutrophils and immunity: From bactericidal action to being conquered. Journal of Immunology Research. 2017:9671604. DOI: 10.1155/2017/9671604 Epub 2017 Feb 19

[31] Li FJ, Starrs L, Burgio G. Tug of war between *Acinetobacter baumannii* and host immune responses. Pathogens and Disease. [Internet]. 2018;**76**(9). Available from: https://academic.oup.com/femspd/article/doi/10.1093/femspd/ftz004/5290314

[32] Lázaro-Díez M, Chapartegui-González I, Redondo-Salvo S, Leigh C, Merino D, Segundo DS, et al. Human neutrophils phagocytose and kill *Acinetobacter baumannii* and *A. pittii*. Scientific Reports. 2017;**7**(1):4571. DOI: 10.1038/s41598-017-04870-8 Erratum in: Sci Rep. 2020;10(1):4797

[33] García-Patiño MG, García-Contreras R, Licona-Limón P. The immune response against *Acinetobacter baumannii*, an emerging pathogen in nosocomial infections. Frontiers in Immunology. 2017;**8**:441. DOI: 10.3389/fimmu.2017.00441 Erratum in: Front Immunol. 2022;13:960356

[34] Jeffreys S, Chambers JP, Yu JJ, Hung CY, Forsthuber T, Arulanandam BP. Insights into *Acinetobacter baumannii* protective immunity. Frontiers in Immunology. 2022;**13**:1070424

[35] Huang W, Zhang Q, Li W, Chen Y, Shu C, Li Q, et al. Anti-outer membrane vesicle antibodies increase antibiotic sensitivity of pan-drug-resistant *Acinetobacter baumannii*. Frontiers in Microbiology. 2019;**10**:1379. DOI: 10.3389/fmicb.2019.01379

[36] Rudenko N, Karatovskaya A, Zamyatina A, Shepelyakovskaya A, Semushina S, Brovko F, et al. Immune response to conjugates of fragments of the type K9 capsular polysaccharide of *Acinetobacter baumannii* with carrier proteins. Microbiology Spectrum. 2022;**10**(5):e01674-e01622

[37] Monem S, Furmanek-Blaszk B, Łupkowska A, Kuczyńska-Wiśnik D, Stojowska-Swędrzyńska K, Laskowska E. Mechanisms protecting *Acinetobacter baumannii* against multiple stresses triggered by the host immune response, antibiotics, and outside-host environment. International Journal of Molecular Sciences. 2020;**21**(15):5498. DOI: 10.3390/ijms21155498

[38] Colquhoun JM, Rather PN. Insights into mechanisms of biofilm formation in *Acinetobacter baumannii* and implications for Uropathogenesis. Frontiers in Cellular and Infection Microbiology. 2020;**10**:253

[39] Morris FC, Dexter C, Kostoulias X, Uddin MI, Peleg AY. The mechanisms of disease caused by *Acinetobacter baumannii*. Frontiers in Microbiology. 2019;**10**:1601

[40] Kang FY, How CK, Wang YC, Cheng A, Yang YS, Kuo SC, et al. Influence of severity of infection on the effect of appropriate antimicrobial therapy for *Acinetobacter baumannii* bacteremic pneumonia. Antimicrobial Resistance and Infection Control. 2020;**9**(1):160

[41] Gu Z, Han Y, Meng T, Zhao S, Zhao X, Gao C, et al. Risk factors and clinical outcomes for patients with *Acinetobacter baumannii* Bacteremia. Medicine (Baltimore). 2016;**95**(9):e2943. DOI: 10.1097/MD.0000000000002943

[42] Qiao L, Zhang MYN, Zhang HZ, Su CL. Analysis of risk factors on the prognosis of *Acinetobacter baumannii* bloodstream infection. Clinical Proteomics & Bioinformatics [Internet]. 2016;**1**(2). DOI: 10.15761/CPB.1000111

[43] McConnell MJ, Actis L, Pachón J. *Acinetobacter baumannii*: Human infections, factors contributing to pathogenesis and animal models. FEMS Microbiology Reviews. 2013;**37**(2):130-155. DOI: 10.1111/j.1574-6976.2012.00344

[44] Kyriakidis I, Vasileiou E, Pana ZD, Tragiannidis A. *Acinetobacter baumannii* antibiotic resistance mechanisms. Pathogens. 2021;**10**(3):373. DOI: 10.3390/pathogens10030373

[45] Sun C, Yu Y, Hua X. Resistance mechanisms of tigecycline in *Acinetobacter baumannii*. Frontiers in Cellular and Infection Microbiology. 2023;**13**:1141490

[46] Nie D, Hu Y, Chen Z, Li M, Hou Z, Luo X, et al. Outer membrane protein a (OmpA) is a potential therapeutic target for *Acinetobacter baumannii* infection. Journal of Biomedical Science. 2020;**27**(1):26

[47] Khalili S, Chen W, Jahangiri A. Editorial: Recent advances in the development of vaccines against *Acinetobacter baumannii*. Frontiers in Immunology. 2023;**14**:1187554

[48] Nielsen TB, Yan J, Slarve M, Lu P, Li R, Ruiz J, et al. Monoclonal antibody therapy against *Acinetobacter baumannii*. Infection and Immunity. 2021;**89**(10):e00162-e00121

Section 2

Urinary tract Infections Caused by *Acinetobacter baumannii*

Chapter 4

Treatment of *Acinetobacter baumannii*

Anup R. Warrier and Sneha Radha

Abstract

Acinetobacter baumannii is a Priority 1 pathogen under the WHO list for research and discovery of new antibiotics. The epidemiology of the pathogen suggests its relevance as an important "healthcare-associated" pathogen—with the most common clinical syndrome being ventilator-associated pneumonia. Rising rates of carbapenem resistance in this pathogen have necessitated re-purposing of old drugs, use of high-dose regimens, and newer antimicrobial options. Combination therapy for carbapenem-resistant isolates, especially in sicker patients, is now advocated. Here, we describe the traditional treatment options and selection of drugs in multidrug-resistant infections, along with a brief review of the evidence followed by emerging treatment options.

Keywords: *Acinetobacter baumannii* infections, multidrug-resistant *Acinetobacter*, gram negative bacterial infections, antimicrobial therapy, carbapenem resistant *Acinetobacter baumanii*

1. Introduction

Acinetobacter baumannii has established itself as an important pathogen over the years, especially in the critical care settings. Often, it has been a pathogen of "intensive care unit (ICU) outbreaks" and a major pathogen for ventilator-associated pneumonia. The varied resistance mechanisms and its potential for environmental persistence have ensured its position as Priority 1 pathogen in the World Health Organization (WHO) list [1]. This necessitates a deeper understanding of the available treatment options, selection of drugs in combination therapy, and emerging treatment options so that we can offer the best chance of survival in the critically ill.

2. Empiric therapy for *A. baumannii* infections

Choosing an empirical cover that includes *A. baumannii* infections depends on various factors, such as the local epidemiology and risk factors in patients, such as mechanical ventilation or long-term hospitalization. An appropriate empiric cover can slash down mortality rates, especially in critical care [2–4]. Addition of antibiotics that cover for carbapenem-resistant *Acinetobacter baumannii* (CRAB) in areas, where there is a higher incidence is recommended [5].

2.1 Epidemiology and local antibiogram

The incidence of *A. baumannii* infection outbreaks can be related to the carbapenem resistance rates of the area. Recent data from a global study on *A. baumannii* has shown around 65% resistance rates to meropenem among clinical isolates [6]. The distribution of CRAB and multidrug-resistant (MDR) Acinetobacter varies between regions in the world, with lower rates in the United States and Central Europe to higher rates in Asia and Africa [7–9]. Within the European subcontinent, the incidence varies as Central Asian and European surveillance of antimicrobial resistance (CAESAR) and European Center for Disease Prevention and Control (ECDC) surveillance report more than 50% of invasive isolates of *A. baumannii* to be carbapenem resistant from Southern and Eastern Europe [10, 11].

A delay in initiation of appropriate antimicrobial therapy can affect the clinical outcome in patients with *A. baumannii* infections, which can be tackled with the help of a hospital-based antibiogram [12–14]. An antibiogram based on overall susceptibility patterns distributed over location and time can suitably guide a clinician in the timely choice of empiric antibiotic [15].

2.2 Risk factors for infection

In high endemic areas of CRAB, certain risk factors can prompt empirical coverage for the same. The most common risk factors include critically ill, prolonged mechanical ventilation, length of hospital or ICU stay, long-term care facility inmates, and previously colonized patients [16–18]. The predilection for colonization in healthcare settings can be explained by the ability of the bacteria to survive in dry surfaces and biofilm production on medical devices, particularly endotracheal tubes. Other risk factors, including malignancy, previous antibiotic use, and re-intubation, among which prior use of antimicrobials, including third-generation cephalosporins and fluoroquinolones, are strong predictors [2, 18].

Multidrug-resistant organism (MDRO) screening or surveillance culture reports are seldom performed in regions with higher prevalence and reported to have lower sensitivity with limited sites of sampling [19]. Treatment or decolonization for MDR Gram-negative organisms based on surveillance sampling is disapproved by many organizations and is only implicated as an infection control measure [20, 21].

Even though nosocomial infection is the dominant picture among Acinetobacter infections, community-acquired infections rarely occur more often in tropical climates, presenting commonly as pneumonia [22–24]. Fulminant infections associated with a high mortality of 64% have also been reported in the Asia-Pacific region [25]. These strains are infrequently resistant to antibiotics but will not be covered by the usual community-acquired pneumonia (CAP) cover, such as ceftriaxone [23, 26].

2.3 Site of infection

Acinetobacter infections can occur in any organ system, with the majority in respiratory tracts causing ventilator-associated pneumonia (VAP) or hospital-acquired pneumonia (HAP). The incidence of VAP varies with the endemicity in a region, comprising an overall incidence rate of 3–7% of VAP/HAP globally, this increases to a maximum of 36% in Asian countries [27–29]. Secondary infections associated with COVID-19 pneumonia are also being reported worldwide [30–32]. Bacteremia and urinary tract infection (UTI) follows, associated with indwelling catheters and immunocompromised conditions [29]. In post-neurosurgical patients with or without intraventricular catheters, *A. baumannii* can cause meningitis, leading to a 70% overall mortality [33, 34]. Skin and soft tissue infections of injuries associated with war and natural disasters reportedly caused by pan-drug isolates and can complicate orthopedic infections [35–37].

2.4 Resistance patterns

Most strains have intrinsic beta-lactamases, providing resistance to penicillin and older generations of cephalosporins. Extended-spectrum beta-lactamases are predominant in many regions, influencing carbapenem over-use, and subsequently breeding multidrug-resistant isolates.

Carbapenem resistance in CRAB is driven chiefly by carbapenemases of classes B (metallo-beta-lactamases) and D (oxacillinases) for which non-beta-lactam choices, such as polymyxins and tetracycline are recommended. Resistance to tigecycline is less compared to minocycline, with most CRAB isolates remaining susceptible and hence can be a good companion drug. Aminoglycoside and fluoroquinolone resistance are common and involve efflux pumps or target modification, conferring high level of resistance to these agents.

Hetero resistance is when subpopulations within a susceptible isolate are resistant but determined as sensitive by standard antimicrobial susceptibility testing (AST) methods, which can lead to clinical failure. This is often seen in cases with previous therapy using colistin and that can possibly be prevented by combination therapy [38–40].

2.5 Choice of empiric therapy

Based on above conditions, an assessment for the need for anti-Acinetobacter cover should be determined as the appropriate and timely initiation of antimicrobial therapy in serious infections is crucial. Also, inappropriate or inadequate empirical antimicrobial choice can lead to increased length of stay, as well as hospital costs [12, 41]. In ICU settings with lower prevalence of CRAB, carbapenems are the drugs of choice. Ertapenem should be avoided as it only has weak action against *Acinetobacter spp.* Combination therapy can be considered in critically ill based on local susceptibility patterns. For mild infections, particularly UTI, monotherapy with cephalosporins or aminoglycosides is a good option with close monitoring of the patient [42].

When there is a higher suspicion of carbapenem resistance, polymyxin-based combination therapy is recommended as empirical therapy. The companion drugs being tetracyclines (tigecycline or minocycline) or sulbactam. Sulbactam in combination with ampicillin and most recently durlobactam has risen as the drug of choice for CRAB infections, but caution is advised for empiric indications due to mounting resistance [5, 43]. The pulmonary endothelial lining fluid (ELF) concentrations are lower for tigecycline with usual dosing for CRAB and are associated with a higher chance for resistance development, hence monotherapy should be avoided [5, 18, 44].

If a second episode of suspected infection occurs when the patient is on an antibiotic for a different infection, it is suggested to choose a different class of antibiotic due to a higher chance of resistance to the ongoing antibiotic [45].

3. Targeted therapy for *A. baumannii* infections

De-escalation or targeting the therapy based on microbiological culture is the recommended step to be taken when culture reports are available as this move can bring in reduction of drug toxicity, unnecessary cost, and prevent antibiotic-associated diarrhea or *Clostridium difficile* infections.

3.1 Colonizers vs. pathogenic *A. baumannii*

Acinetobacter baumannii is acquired through the hospital environment from surfaces and hands of healthcare workers. It commonly colonizes the respiratory tract, skin, and any indwelling catheters of a patient. Hence, sampling can often detect such colonizing organisms, which need to be differentiated from infection. The task becomes more perplexing yet key in immunocompromised or severely ill [46]. This is more relevant in CRAB isolates as differentiating plays a decisive role on need for expensive and restricted antimicrobial agents for treatment.

Clinical, radiological, and laboratory parameters can aid in differentiating colonization from infection. When isolated from sterile sites, such as blood culture or cerebrospinal fluid culture, treatment is mandatory. Treatment includes both the removal of indwelling catheter if present and appropriate antimicrobial. Antibiotics are not recommended when the culture is positive from a non-sterile site from a patient with no signs of infection. Parameters that help in diagnosing infection by *A. baumannii* are admission to ICU, number of days of hospitalization, absolute neutrophil count (ANC), and C-reactive protein (CRP) according to two prospective cohort studies [47, 48]. Clinical-pulmonary infection score (CPIS) score developed for diagnosing VAP using fever, endotracheal tube (ET) secretions, leukocytosis, PaO2/FiO2, chest radiographic picture, and isolation of pathogenic bacteria in culture is ideally used for determining the need for bronchoalveolar lavage (BAL) but can also confirm the presence of pneumonia and relevance of culture.

3.2 Non-carbapenem-resistant *A. baumannii*

3.2.1 Beta lactams

Most clinical isolates have intrinsic beta-lactamase production that lyses penicillin and first-generation cephalosporins. If susceptible to penicillin, these agents are the drugs of choice for Acinetobacter infections, including third-generation cephalosporins. In the presence of extended-spectrum Beta-lactamase (ESBLs) and Acinetobacter-derived cephalosporins (ADCs), carbapenems become the agent of choice, with ertapenem having a weak activity against *Acinetobacter spp* [49]. These agents are ideal for its bactericidal action and good pharmacotherapeutic properties.

3.2.2 Beta-lactamase inhibitors

All beta-lactamase inhibitors, such as clavulanate and tazobactam, have intrinsic activity against Acinetobacter, but sulbactam has the better activity among them [50]. For sensitive isolates with MIC <4 mg/L, sulbactam at a lower dose of 4 grams per day is sufficient to be infused in 350 ml normal saline over 4 hours. Most commonly, this is available as the formulation of ampicillin sulbactam or cefoperazone sulbactam.

There is a rising MIC trend for sulbactam that impedes its use as empiric therapy or monotherapy for severe infections [43].

3.2.3 Aminoglycosides

Amikacin and tobramycin are the most active agents in the group. These agents have low lung and CSF penetration with high toxicity profiles. With higher chances of bacteriological failures and dosing concerns in critically ill, aminoglycosides are not recommended as monotherapy except for urinary tract infections, where they reach in very high concentrations [51, 52]. Amikacin and gentamicin can be administered intrathecally or intraventricular for CRAB meningitis or ventriculitis.

3.2.4 Quinolones

Susceptibility to these agents is lower, and thus is less commonly used in the treatment of these infections. Due to its good pharmacotherapeutic properties and oral bioavailability, quinolones are a good option if susceptible. Resistance to these agents arises along with other antimicrobials with multiple mechanisms mainly arising with mutations in *gyrA* and *parC* genes.

3.3 Carbapenem-resistant *A. baumannii*

Most *A. baumannii* infections are caused by carbapenem-resistant strains in nosocomial settings due to the capacity of the organism to acquire resistance genes and its resilience in the hospital environment. Mortality associated with MDR *A. baumanii* strains is higher than in susceptible organisms [53].

3.3.1 Sulbactam

A penicillanic acid derivative that has intrinsic activity against Acinetobacter by saturating PBPs 1, 2, and 3, especially with higher doses [54]. The beta-lactamases produced by CRAB can lyse sulbactam, which is observed *in vitro* and reflected in the international surveillance systems, such as The Clinical and Laboratory Standards Institute (CLSI) [55]. But this was not observed in clinical trials, where sulbactam activity is intact even for MIC>16 mg/L, when given as 9 gram/day dosing over 4 hours infusion [56]. At this dose, sulbactam overcomes resistance by Oxa-23 beta-lactamases and has shown effectiveness more with meropenem [42, 57, 58]. At lower MICs of sulbactam, lower doses of 1 g sulbactam every fourth to sixth hourly should be enough. Ampicillin sulbactam is the commonest formulation available for sulbactam with 2:1 ratio of ampicillin to sulbactam that is recommended by Infectious diseases society of America (IDSA) and European Society of Clinical Microbiology and Infectious Diseases (ESCMID) guidelines [59, 60]. For mild invasive CRAB infections, monotherapy with ampicillin sulbactam is the treatment of choice and is given as 27 g per day dosing over 4 hours [59].

When compared to colistin, previously, the first choice for CRAB infections, sulbactam has better kinetics and lesser nephrotoxicity, thus showing improved clinical outcomes in research [61–63]. This brought sulbactam to the limelight, even with conflicting issues on standardized susceptibility testing for CRAB isolates and studies showing lower mortality with colistin therapy [64]. High doses of sulbactam can be associated with hepatotoxicity, and it is suggested to monitor LFT while on therapy.

The resistance to sulbactam arises through reduced expression of PBP 2 and increased TEM-1 expression [65, 66]. Hence, it warrants monitoring for resistance development and the use of combination therapy for severe and complicated infections.

3.3.2 Polymyxins

Polymyxins are polypeptide cationic antibiotics comprised of polymyxin B and E (colistin). Colistin is the most widely used form worldwide, which is available in its prodrug form as colistimethate (CMS). Polymyxins were once considered the main backbone for CRAB infections but were replaced by sulbactam in the recent IDSA and ESCMID guidance documents. Current recommendations by IDSA recommend polymyxin B as the preferred form except in UTI and to be given in combination with other agents according to susceptibility patterns [59]. Due to a high risk of nephrotoxicity with colistin [63] and the poor ELF concentrations in critically ill, polymyxin B is the preferred choice except in UTI, where colistin achieves better concentrations.

Dosing of polymyxin B is body weight based with a loading dose and given when MIC is less than 2 mg/L for CRAB isolates. A loading dose of colistin, which is needed to achieve target plasma concentration, has been shown to increase mortality when administered to critically ill patients with CRAB infections [67, 68]. Resistance develops by modification of LPS, the target site of polymyxin action, through plasmid acquired resistance genes. Intrathecal colistin and polymyxin B can be administered for CRAB meningitis or ventriculitis with systemic antimicrobial. The dosing varies for both polymyxins ranging from 20,000 IU to 250,000 IU per day for polymyxin B to 5–20 mg per day for colistin [34, 69, 70].

3.3.3 Tetracyclines

Minocycline and tigecycline are active against CRAB even when other tetracyclines are found resistant, with tigecycline having a more than 90% susceptibility among CRAB isolates based on a countrywide surveillance in Europe [43]. Due to its bactericidal nature and unclear pharmacokinetics, both are suggested to be given in combination therapy for CRAB treatment. Minocycline is available both orally and parentally, whereas tigecycline is available only as IV formulation.

The breakpoint for tigecycline in CRAB is not recommended by CLSI or EUCAST and a (food and drug administration) FDA approved clinical breakpoint of 2 mg/l for *Enterobacterales* is adopted. With MIC <2 mg/l, combination therapy has shown benefit, even though earlier studies have shown higher all-cause mortality in the critically ill [71–74]. These studies have focused mostly on bacteremia and pneumonia, where tigecycline concentrations are very low, that is. plasma and ELF concentrations, respectively. A higher dose of tigecycline with 200 mg of loading dose followed by 100 mg twice daily doses depicts a better outcome in MDR GNB infections [75, 76] and is recommended. The PK parameters for minocycline were achieved better for CRAB pneumonia with higher doses of 200 mg twice daily [77].

3.4 Monotherapy vs. combination therapy

The rationale for combination therapy is built on the concept of various *in vitro* synergism, pharmacotherapeutic advantage of overcoming the poor pharmacodynamics of individual agents, unfavorable clinical outcomes of invasive CRAB infections, and the higher possibility in prevention of emerging resistance [78, 79].

Several studies demonstrate better clinical cure with combination therapies, most often containing colistin and others have shown higher microbiological cure rates. A Bayesian analysis that included 23 studies compared colistin monotherapy with combination regimens, which depicted sulbactam to have the highest survival benefit among critically ill [71].

3.4.1 Colistin with meropenem

This was an advocated as a common modality earlier based on *in vitro* experiments indicating synergy and reduced bacterial growth with the combination [80]. But disproved clinically, by two randomized control trials in ICU patients that showed no difference between the clinical cure rate or 28–day mortality rates between colistin monotherapy and colistin with meropenem combination [81, 82]. The *in vitro* benefits could not be translated clinically with these studies as colistin levels in ELF are lower, especially in the critically ill, and the isolates in these studies had a high level of carbapenem resistance [79].

3.4.2 Three-drug combination

With the addition of ampicillin sulbactam to the above combination, the results show significant improvement in terms of 30-day mortality also, with one of the recent study reporting on a likely suppression of resistance emergence among COVID-19 patients with CRAB infections [83, 84]. Another triple therapy consisting of colistin, tigecycline, and sulbactam showed the highest clinical cure rate among various treatment options for MDR and extremely drug-resistant (XDR) *A. baumannii* infections [64]. Thus, triple therapy is a suggested approach for extremely resistant CRAB than other dual combinations [78, 79].

3.4.3 Polymyxin-based combination therapy

The main researched combinations included colistin combinations with either sulbactam, tigecycline, fosfomycin, or rifampicin, which have mixed evidence in terms of clinical cure and microbiological cure. A meta-analysis on 29 studies consisting of over 2000 patients revealed a higher microbiological cure for the sulbactam—colistin combination when compared to colistin with tigecycline or colistin alone [64]. *In vitro* synergy testing by checkerboard method and time-kill analysis indicates the highest synergy between minocycline and colistin [85, 86]. Whereas, colistin with tigecycline showed an antagonism, but such inhibition is absent in clinical trials, where it has shown better microbiological cure rates but not improved mortality benefits [72, 87, 88].

Combinations with rifampicin and fosfomycin have a good microbiological response even within 72 hours of therapy, but there is no evidence of significant difference in mortality from monotherapy [89, 90]. The combination of rifampicin and tigecycline with colistin, respectively, has a good anti-biofilm action that can be used effectively for antibiotic lock therapy [91, 92]. Notably, most studies on polymyxin combinations included colistin and nephrotoxicity was an associated adverse drug event (ADE), and this ADE could have been easily avoided with the use of polymyxin B, an active form of colistin, which not only has a lower risk of nephrotoxicity but also has better steady-state concentrations in plasma. Only few studies based on polymyxin B have been conducted and a combination with this agent is preferred with reduced mortality among critically ill [59, 60, 93, 94].

3.4.4 Sulbactam combination therapies

Sulbactam combinations are the mainstay for moderate to severe invasive CRAB infections, as stated by the latest IDSA AMR guidance document and endorsed by the ESCMID 2022 guidelines. Combination of high-dose sulbactam with tigecycline and quinolone has shown the best clinical outcome, but the numbers are less compared to colistin-based regimens [95, 96]. There is a need for focus on research of such combination therapy for the better understanding of dosing and efficacy in critically ill.

3.4.5 Other combinations

Strong synergy exists between cefiderocol and meropenem as shown by an *in vitro* study but has not been clinically evaluated [97]. Combination of colistin with glycopeptides has been studied, but combination with vancomycin has shown high nephrotoxicity [98, 99]. Such combinations do not have enough supporting data and is to be avoided.

However, failure with combination therapy has been shown in patients with sepsis. A metanalysis on drug-resistant *A. baumannii* showed only three studies to depict a superiority of combination vs. monotherapy from a total of 12 studies. The concern with combination therapy arises with the associated increased cost and toxicity from multiple antimicrobial agents used. The risk of *C. difficile* can also increase when inciting antibiotics are given for treatment [78]. Thus, de-escalation to susceptible agents based on culture reports when available is advised in mild cases of invasive CRAB infections [5, 59].

3.5 Role of inhaled antimicrobials

Aminoglycosides and colistin are often nebulized for patients with MDR gram-negative bacterial (GNB) VAP or tracheobronchitis. The role of nebulized antibiotics is controversial, with IDSA against its administration with or without IV antimicrobials. Studies using high dose (5 million units twice daily) colistin with vibrating nebulizer along with intravenous (IV) antimicrobial agent have shown benefits [100]. For tracheobronchitis and when susceptible in non-resolving cases of CRAB pneumonia, nebulization can be attempted as an adjunctive therapy (**Figure 1**).

3.6 Duration of therapy

There is a lack of consensus for specific duration for MDR infections and in particular CRAB. Studies on patients with MDR infections with VAP and BSI have used a range of 7–22 days of therapy [101–103]. RCT and prospective studies on VAP have shown no difference in mortality with shorter courses of 3–8 days [104–106]. But few studies show an occurrence of relapses in few patients following short courses for gram-negative pathogens are of concern according to some [106, 107].

Duration of empiric therapy for CRAB, where cultures are negative or limited resources for diagnostics, should be based on site and severity of infection and discontinued if an alternative diagnosis is confirmed. For carbapenem-sensitive AB infection therapy, duration is based on site and severity of infection and longer duration of therapy required for meningitis and joint infections. Whereas, a longer duration is suggested for severe CRAB infections with a minimum of 14 days of therapy and even longer, up to 4 to 8 weeks in the presence of complicated infections, such as post-neurosurgical meningitis/ventriculitis or joint infections. In meningitis

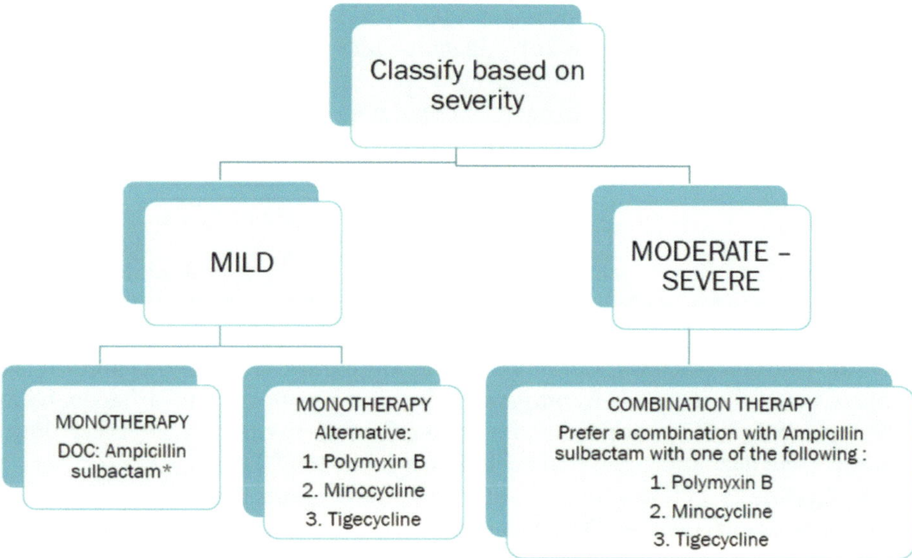

Figure 1.
*Algorithm for treatment of invasive CRAB infections based on IDSA/ESCMID guidelines. *Ampicillin sulbactam is the current fixed drug combination that is available widely.*

or ventriculitis, the duration of intraventricular antimicrobial will depend on three negative CSF samples according to the IDSA recommendations [108].

4. Novel treatment strategies

4.1 Newer antimicrobial agents

Antimicrobial options are limited for treatment of CRAB, with resistance developing rapidly in the organism. Newer options are sought to overcome the glitches in the treatment options that are available currently such as toxicity, pharmacokinetic issues, emergence of resistance, and availability.

4.1.1 Cefiderocol

A siderophore cephalosporin with a wide range of beta-lactamase resistance, and hence suggested for CRE, CRPA, and CRAB. The FDA approved its use in complicated UTI, including pyelonephritis and hospital-acquired pneumonia. In spite of the promising *in vitro* actions of the drug, two large studies found no significant difference in mortality compared to colistin-based therapies for CRAB [109–111] and higher mortality rates with monotherapy [112]. Also, the Italian study discovered four out of eight isolates from microbiologically failed cases to be cefiderocol resistant [109]. This led the IDSA to suggest cefiderocol as a last resort and only to be given as combination therapy for CRAB infections [59].

4.1.2 Durlobactam: Sulbactam

This is the most recently FDA-approved agent for treatment of CRAB pneumonia [113]. Durlobactam is a next generation diazabicyclooctanone (DBO) beta-lactamase inhibitor that is resistant to lyses by Class A, C, and D oxacillinases. Combined with sulbactam, it potentiates the action of sulbactam against CRAB up to 32-fold of MIC [55]. A phase three trials on CRAB pneumonia patients on either sulbactam durlobactam or colistin with a combination agent showed non-inferiority in terms of 28-day mortality and lower adverse events [114].

4.1.3 Eravacycline

A novel synthetic fluorocycline, such as tigecycline, displays good *in vitro* action against MDR pathogens, including GNB and GPC microorganisms. It has a lower MIC in CRAB than tigecycline or minocycline with reliable *in vitro* activity against oxacillinases and colistin-resistant isolates [115]. Nevertheless, the clinical trials on UTI and IAI show non-inferiority of the drug when compared to inactive agents, such as carbapenems and quinolones [116–118]. The proportion of CRAB infections in these studies is very low, and thus, we will need more research on its *in vivo* action on CRAB.

Omadacycline and plazomicin are some other new agents that are active on CRAB isolates. A newer tetracycline, Omadacycline, has a spectrum of activity similar to minocycline and has action against CRAB isolates [119, 120]. Whereas, plazomicin is a next generation aminoglycoside with extended spectrum, including CRAB. This drug is approved by FDA for the treatment of carbapenem-resistant Enterobacterales but has shown promising activity against CRAB, including in combination with other drugs [121, 122].

4.2 Other therapeutic options

Bacteriophage, antimicrobial peptides (AMP), immunotherapy, monoclonal antibodies, and endolysin are some of the potential non-antimicrobial agents that are being extensively researched [123]. Phage-related therapy is unique in the sense that it is highly specific to the targeted pathogens and have lesser toxicity. But the clinical efficacy associated with such therapies are yet to be demonstrated. Phage SH-Ab15519 and Acinetobacter phage Bϕ-R2096 are novel Acinetobacter phages, which are considered safe based on genomic studies [124, 125]. Phage-antibiotic combinations based on a phenomenon termed phage-antibiotic synergy has been exhibited in *A. baumannii* on colistin MIC [126] and also depicted in human trial [127]. AMP formed from other living organisms as part of their innate immune mechanisms can be used against infections as an adjunctive therapy. This has an advantage of lower chances of resistance development [128].

5. Conclusion

Acinetobacter baumannii remains a "high priority" pathogen and of great clinical significance, especially in the critically ill ICU patient. With a significant proportion of the isolates demonstrating resistance to traditional "drugs of choice," such as carbapenems, we have moved on to repurposed older drugs—polymyxins and high-dose Sulbactam—as primary drugs for treating serious infections. Tetracyclines—old tigecycline, minocycline at "double dose" and new (Eravacycline and Omadacycline) have been the next plausible treatment options. We also fall back upon combination therapy with older drugs/with or without the newer options for pan-drug-resistant isolates. Drugs, such as cefiderocol and sulbactam-durlobactam, hold promise for the future. However, the identification or differentiation of a patient colonized with *Acinetobacter baumannii* versus true invasive infection/disease constitutes the most important treatment decision.

Conflict of interest

The authors declare no conflict of interest.

Appendices and nomenclature

ADC	*Acinetobacter*-derived cephalosporinases
ADE	adverse drug event
ANC	absolute neutrophil count
AMR	antimicrobial resistance
BSI	infection
CAESAR	Central Asian and European surveillance of antimicrobial resistance
CPIS	clinical-pulmonary infection score
CRAB	carbapenem-resistant *Acinetobacter baumannii*
CRE	carbapenem-resistant enterobacterales
CRP	C-reactive protein
CRPA	carbapenem-resistant *pseudomonas aeruginosa*
CSF	cerebrospinal fluid
CLSI	The clinical and laboratory standards institute
ECDC	European centre for disease prevention and control
ELF	endothelial lung fluid
ESBL	extended spectrum Beta-lactamase
ESCMID	European society of clinical microbiology and infectious diseases
ET	endotracheal secretions
EUCAST	European society of clinical microbiology and infectious diseases
FDA	food and drug administration
FiO2	fraction of inspired oxygen
HAP	hospital-acquired pneumonia
ICU	intensive care unit
IDSA	infectious diseases society of America
IV	intravenous
LFT	liver function test
LPS	lipopolysaccharide
MDR	multidrug resistant
MDRO	multidrug-resistant organism
MIC	minimum inhibitory concentration
PaO2	partial pressure of oxygen in arterial blood
PBP	penicillin-binding protein
PK	pharmacokinetic
RCT	randomized controlled trial
TEM	Class A beta-lactamase first isolated from a patient called Temoneira
UTI	urinary tract infection
VAP	ventilator-associated pneumonia
XDR	extremely drug resistant

Author details

Anup R. Warrier* and Sneha Radha
Aster Medcity, Kochi, India

*Address all correspondence to: dranup.warrier@asterdmhealthcare.com

References

[1] WHO Publishes List of Bacteria for Which New Antibiotics are Urgently Needed [Internet]. 2017. Available from: https://www.who.int/news/item/27-02-2017-who-publishes-list-of-bacteria-for-which-new-antibiotics-are-urgently-needed [Accessed: August 29, 2023]

[2] Garnacho-Montero J, Ortiz-Leyba C, Fernández-Hinojosa E, Aldabó-Pallás T, Cayuela A, Marquez-Vácaro JA, et al. *Acinetobacter baumannii* ventilator-associated pneumonia: Epidemiological and clinical findings. Intensive Care Medicine. 2005;**31**(5):649-655. DOI: 10.1007/s00134-005-2598-0

[3] Cisneros JM, Reyes MJ, Pachón J, Becerril B, Caballero FJ, García Garmendia JL, et al. Bacteremia due to *Acinetobacter baumannii*: Epidemiology, clinical findings, and prognostic features. Clinical Infectious Diseases. 1996;**22**(6):1026-1032. DOI: 10.1093/clinids/22.6.1026

[4] Vincent JL, Rello J, Marshall J, Silva E, Anzueto A, Martin CD, et al. International study of the prevalence and outcomes of infection in intensive care units. Journal of the American Medical Association. 2009;**302**(21):2323-2329. DOI: 10.1001/jama.2009.1754

[5] Garnacho-Montero J, Dimopoulos G, Poulakou G, Akova M, Cisneros JM, De Waele J, et al. Task force on management and prevention of *Acinetobacter baumannii* infections in the ICU. Intensive Care Medicine. 2015;**41**(12):2057-2075. DOI: 10.1007/s00134-015-4079-4

[6] Seifert H, Blondeau J, Lucaßen K, Utt EA. Global update on the in vitro activity of tigecycline and comparators against isolates of *Acinetobacter baumannii* and rates of resistant phenotypes (2016-2018). Journal of Global Antimicrobial Resistance. 2022;**31**:82-89. DOI: 10.1016/j.jgar.2022.08.002

[7] Giammanco A, Calà C, Fasciana T, Dowzicky MJ. Global assessment of the activity of Tigecycline against multidrug-resistant gram-negative pathogens between 2004 and 2014 as part of the Tigecycline evaluation and surveillance trial. mSphere. 2017;**2**(1):e00310-e00316. DOI: 10.1128/mSphere.00310-16

[8] CDC. The Biggest Antibiotic-Resistant Threats in the U.S. Centers for Disease Control and Prevention. [Internet]. 2022. Available from: https://www.cdc.gov/drugresistance/biggest-threats.html [Accessed: August 20, 2023]

[9] Castanheira M, Mendes RE, Gales AC. Global epidemiology and mechanisms of resistance of *Acinetobacter baumannii-calcoaceticus* complex. Clinical Infectious Diseases. 2023;**76**(Suppl 2):S166-S178. DOI: 10.1093/cid/ciad109

[10] World Health Organization. Regional Office for Europe. Central Asian and European Surveillance of Antimicrobial Resistance: Annual Report 2020 [Internet]. Available from: https://apps.who.int/iris/handle/10665/345873 [Accessed: August 20, 2023]

[11] Antimicrobial Resistance Surveillance in Europe 2022-2020 Data [Internet]. 2022. Available from: https://www.ecdc.europa.eu/en/publications-data/antimicrobial-resistance-surveillance-europe-2022-2020-data [Accessed: August 20, 2023]

[12] Erbay A, İdil A, Gözel MG, Mumcuoğlu İ, Balaban N. Impact of

early appropriate antimicrobial therapy on survival in *Acinetobacter baumannii* bloodstream infections. International Journal of Antimicrobial Agents. 2009;**34**(6):575-579. DOI: 10.1016/j.ijantimicag.2009.07.006

[13] Grupper M, Sprecher H, Mashiach T, Finkelstein R. Attributable mortality of nosocomial Acinetobacter bacteremia. Infection Control and Hospital Epidemiology. 2007;**28**(3):293-298. DOI: 10.1086/512629

[14] Zak-Doron Y, Dishon Benattar Y, Pfeffer I, Daikos GL, Skiada A, Antoniadou A, et al. The association between empirical antibiotic treatment and mortality in severe infections caused by Carbapenem-resistant gram-negative bacteria: A prospective study. Clinical Infectious Diseases. 2018;**67**(12):1815-1823. DOI: 10.1093/cid/ciy371

[15] Wenzler E, Goff DA, Humphries R, Goldstein EJC. Anticipating the unpredictable: A review of antimicrobial stewardship and Acinetobacter infections. Infectious Disease and Therapy. 2017;**6**(2):149-172. DOI: 10.1007/s40121-017-0149-y

[16] Fournier PE, Richet H, Weinstein RA. The epidemiology and control of *Acinetobacter baumannii* in health care facilities. Clinical Infectious Diseases. 2006;**42**(5):692-699. DOI: 10.1086/500202

[17] Huang H, Chen B, Liu G, Ran J, Lian X, Huang X, et al. A multi-center study on the risk factors of infection caused by multi-drug resistant *Acinetobacter baumannii*. BMC Infectious Diseases. 2018;**18**(1):11. DOI: 10.1186/s12879-017-2932-5

[18] Garnacho-Montero J, Timsit JF. Managing *Acinetobacter baumannii* infections. Current Opinion in Infectious Diseases. 2019;**32**(1):69. DOI: 10.1097/QCO.0000000000000518

[19] Marchaim D, Navon-Venezia S, Schwartz D, Tarabeia J, Fefer I, Schwaber MJ, et al. Surveillance cultures and duration of carriage of multidrug-resistant *Acinetobacter baumannii*. Journal of Clinical Microbiology. 2007;**45**(5):1551-1555. DOI: 10.1128/JCM.02424-06

[20] Tacconelli E, Mazzaferri F, de Smet AM, Bragantini D, Eggimann P, Huttner BD, et al. ESCMID-EUCIC clinical guidelines on decolonization of multidrug-resistant gram-negative bacteria carriers. Clinical Microbiology and Infection. 2019;**25**(7):807-817. DOI: 10.1016/j.cmi.2019.01.005

[21] Verdugo-Paiva F, Otaiza F, Roson-Rodríguez P, Rojas-Gomez AM, Galas M, Omeiri NE, et al. Effects of screening strategies to detect carbapenem-resistant gram-negative bacteria: A systematic review. American Journal of Infection Control. 2022;**50**(12):1381-1388. DOI: 10.1016/j.ajic.2022.02.018

[22] Dexter C, Murray GL, Paulsen IT, Peleg AY. Community-acquired *Acinetobacter baumannii*: Clinical characteristics, epidemiology and pathogenesis. Expert Review of Anti-Infective Therapy. 2015;**13**(5):567-573. DOI: 10.1586/14787210.2015.1025055

[23] Falagas ME, Karveli EA, Kelesidis I, Kelesidis T. Community-acquired Acinetobacter infections. European Journal of Clinical Microbiology & Infectious Diseases. 2007;**26**(12):857-868. DOI: 10.1007/s10096-007-0365-6

[24] Anstey NM, Currie BJ, Hassell M, Palmer D, Dwyer B,

SeifertH. Community-acquired Bacteremic Acinetobacter pneumonia in tropical Australia is caused by diverse strains of *Acinetobacter baumannii*, with carriage in the throat in At-risk groups. Journal of Clinical Microbiology. 2002;**40**(2):685-686. DOI: 10.1128/JCM.40.2.685-686.2002

[25] Leung WS, Chu CM, Tsang KY, Lo FH, Lo KF, Ho PL. Fulminant community-acquired *Acinetobacter baumannii* pneumonia as a distinct clinical syndrome. Chest. 2006;**129**(1):102-109. DOI: 10.1378/chest.129.1.102

[26] Davis JS, McMillan M, Swaminathan A, Kelly JA, Piera KE, Baird RW, et al. A 16-year prospective study of community-onset Bacteremic Acinetobacter pneumonia: Low mortality with appropriate initial empirical antibiotic protocols. Chest. 2014;**146**(4):1038-1045. DOI: 10.1378/chest.13-3065

[27] Chung DR, Song JH, Kim SH, Thamlikitkul V, Huang SG, Wang H, et al. High prevalence of multidrug-resistant nonfermenters in hospital-acquired pneumonia in Asia. American Journal of Respiratory and Critical Care Medicine. 2011;**184**(12):1409-1417. DOI: 10.1164/rccm.201102-0349OC

[28] Lynch JP, Zhanel GG, Clark NM. Infections due to *Acinetobacter baumannii* in the ICU: Treatment options. Seminars in Respiratory and Critical Care Medicine. 2017;**38**(3):311-325. DOI: 10.1055/s-0037-1599225

[29] Ma C, McClean S. Mapping global prevalence of *Acinetobacter baumannii* and recent vaccine development to tackle it. Vaccine. 2021;**9**(6):570. DOI: 10.3390/vaccines9060570

[30] Perez S, Innes GK, Walters MS, Mehr J, Arias J, Greeley R, et al. Increase in hospital-acquired Carbapenem-resistant *Acinetobacter baumannii* infection and colonization in an acute care hospital during a surge in COVID-19 admissions—New Jersey, February-July 2020. MMWR. Morbidity and Mortality Weekly Report. 2020;**69**(48):1827-1831. DOI: 10.15585/mmwr.mm6948e1

[31] Sharifipour E, Shams S, Esmkhani M, Khodadadi J, Fotouhi-Ardakani R, Koohpaei A, et al. Evaluation of bacterial co-infections of the respiratory tract in COVID-19 patients admitted to ICU. BMC Infectious Diseases. 2020;**20**(1):646. DOI: 10.1186/s12879-020-05374-z

[32] Chen X, Liao B, Cheng L, Peng X, Xu X, Li Y, et al. The microbial coinfection in COVID-19. Applied Microbiology and Biotechnology. 2020;**104**(18):7777-7785. DOI: 10.1007/s00253-020-10814-6

[33] Metan G, Alp E, Aygen B, Sumerkan B. *Acinetobacter baumannii* meningitis in post-neurosurgical patients: Clinical outcome and impact of carbapenem resistance. The Journal of Antimicrobial Chemotherapy. 2007;**60**(1):197-199. DOI: 10.1093/jac/dkm181

[34] Rodríguez Guardado A, Blanco A, Asensi V, Pérez F, Rial JC, Pintado V, et al. Multidrug-resistant Acinetobacter meningitis in neurosurgical patients with intraventricular catheters: Assessment of different treatments. The Journal of Antimicrobial Chemotherapy. 2008;**61**(4):908-913. DOI: 10.1093/jac/dkn018

[35] Rafei R, Dabboussi F, Hamze M, Eveillard M, Lemarié C, Mallat H, et al. First report of blaNDM-1-producing *Acinetobacter baumannii* isolated in Lebanon from civilians wounded during the Syrian war. International Journal

of Infectious Diseases. 2014;**21**:21-23. DOI: 10.1016/j.ijid.2014.01.004

[36] Tao C, Kang M, Chen Z, Xie Y, Fan H, Qin L, et al. Microbiologic study of the pathogens isolated from wound culture among Wenchuan earthquake survivors. Diagnostic Microbiology and Infectious Disease. 2009;**63**(3):268-270. DOI: 10.1016/j.diagmicrobio.2008.11.009

[37] Guerrero DM, Perez F, Conger NG, Solomkin JS, Adams MD, Rather PN, et al. *Acinetobacter baumannii*-associated skin and soft tissue infections: Recognizing a broadening Spectrum of disease*. Surgical Infections. 2010;**11**(1):49-57. DOI: 10.1089/sur.2009.022

[38] Hong YK, Kim H, Ko KS. Two types of colistin heteroresistance in *Acinetobacter baumannii* isolates. Emerging Microbes & Infections. 2020;**9**(1):2114-2123. DOI: 10.1080/22221751.2020.1821584

[39] Karakonstantis S, Saridakis I. Colistin heteroresistance in *Acinetobacter spp*.: Systematic review and meta-analysis of the prevalence and discussion of the mechanisms and potential therapeutic implications. International Journal of Antimicrobial Agents. 2020;**56**(2):106065. DOI: 10.1016/j.ijantimicag.2020.106065

[40] Yau W, Owen RJ, Poudyal A, Bell JM, Turnidge JD, Yu HH, et al. Colistin hetero-resistance in multidrug-resistant *Acinetobacter baumannii* clinical isolates from the Western Pacific region in the SENTRY antimicrobial surveillance programme. The Journal of Infection. 2009;**58**(2):138-144. DOI: 10.1016/j.jinf.2008.11.002

[41] Dupont H, Mentec H, Sollet JP, Bleichner G. Impact of appropriateness of initial antibiotic therapy on the outcome of ventilator-associated pneumonia. Intensive Care Medicine. 2001;**27**(2):355-362. DOI: 10.1007/s001340000640

[42] Abdul-Mutakabbir JC, Griffith NC, Shields RK, Tverdek FP, Escobar ZK. Contemporary perspective on the treatment of *Acinetobacter baumannii* infections: Insights from the Society of Infectious Diseases Pharmacists. Infectious Disease and Therapy. 2021;**10**(4):2177-2202. DOI: 10.1007/s40121-021-00541-4

[43] Jones RN, Flonta M, Gurler N, Cepparulo M, Mendes RE, Castanheira M. Resistance surveillance program report for selected European nations (2011). Diagnostic Microbiology and Infectious Disease. 2014;**78**(4):429-436. DOI: 10.1016/j.diagmicrobio.2013.10.008

[44] Burkhardt O, Rauch K, Kaever V, Hadem J, Kielstein JT, Welte T. Tigecycline possibly underdosed for the treatment of pneumonia: A pharmacokinetic viewpoint. International Journal of Antimicrobial Agents. 2009;**34**(1):101-102. DOI: 10.1016/j.ijantimicag.2009.01.015

[45] Guidelines for the Management of Adults with hospital-acquired, ventilator-associated, and healthcare-associated pneumonia. American Thoracic Society, Infectious Diseases Society of America. American Journal of Respiratory and Critical Care Medicine. 2005;**171**(4):388-416. DOI: 10.1164/rccm.200405-644ST

[46] Bartal C, Rolston KVI, Nesher L. Carbapenem-resistant *Acinetobacter baumannii*: Colonization, infection and current treatment options. Infectious Disease and Therapy. 2022;**11**(2):683-694. DOI: 10.1007/s40121-022-00597-w

[47] Martín-Aspas A, Guerrero-Sánchez FM, García-Colchero F, Rodríguez-Roca S, Girón-González JA. Differential characteristics of *Acinetobacter baumannii* colonization and infection: Risk factors, clinical picture, and mortality. Infection and Drug Resistance. 2018;**11**:861-872. DOI: 10.2147/IDR.S163944

[48] Feng DY, Zhou JX, Li X, Wu WB, Zhou YQ, Zhang TT. Differentiation between *Acinetobacter baumannii* colonization and infection and the clinical outcome prediction by infection in lower respiratory tract. Infection and Drug Resistance. 2022;**15**:5401-5409. DOI: 10.2147/IDR.S377480

[49] Livermore DM, Sefton AM, Scott GM. Properties and potential of ertapenem. The Journal of Antimicrobial Chemotherapy. 2003;**52**(3):331-344. DOI: 10.1093/jac/dkg375

[50] Higgins PG, Wisplinghoff H, Stefanik D, Seifert H. In vitro activities of the beta-lactamase inhibitors clavulanic acid, sulbactam, and tazobactam alone or in combination with beta-lactams against epidemiologically characterized multidrug-resistant *Acinetobacter baumannii* strains. Antimicrobial Agents and Chemotherapy. 2004;**48**(5):1586-1592. DOI: 10.1128/AAC.48.5.1586-1592.2004

[51] Vidal L, Gafter-Gvili A, Borok S, Fraser A, Leibovici L, Paul M. Efficacy and safety of aminoglycoside monotherapy: Systematic review and meta-analysis of randomized controlled trials. The Journal of Antimicrobial Chemotherapy. 2007;**60**(2):247-257. DOI: 10.1093/jac/dkm193

[52] Thy M, Timsit JF, de Montmollin E. Aminoglycosides for the treatment of severe infection due to resistant gram-negative pathogens. Antibiotics. 2023;**12**(5):860. DOI: 10.3390/antibiotics12050860

[53] Brahmi N, Beji O, Abidi N, Kouraichi N, Blel Y, Ghord HE, et al. Epidemiology and risk factors for colonization and infection by *Acinetobacter baumannii* in an ICU in Tunisia, where this pathogen is endemic. Journal of Infection and Chemotherapy. 2007;**13**(6):400-404. DOI: 10.1007/s10156-007-0557-0

[54] Levin AS. Multiresistant Acinetobacter infections: A role for sulbactam combinations in overcoming an emerging worldwide problem. Clinical Microbiology and Infection. 2002;**8**(3):144-153. DOI: 10.1046/j.1469-0691.2002.00415.x

[55] Karlowsky JA, Hackel MA, McLeod SM, Miller AA. In vitro activity of Sulbactam-Durlobactam against global isolates of *Acinetobacter baumannii-calcoaceticus* complex collected from 2016 to 2021. Antimicrobial Agents and Chemotherapy. 2022;**66**(9):e0078122. DOI: 10.1128/aac.00781-22

[56] Abouelhassan Y, Kuti JL, Nicolau DP, Abdelraouf K. Sulbactam against *Acinetobacter baumannii* pneumonia: Pharmacokinetic/pharmacodynamic appraisal of current dosing recommendations. Open Forum Infectious Diseases. 1 Dec 2022;**9**(2):ofac492-118. DOI: 10.1093/ofid/ofac492.118

[57] Hiraki Y, Yoshida M, Masuda Y, Inoue D, Tsuji Y, Kamimura H, et al. Successful treatment of skin and soft tissue infection due to carbapenem-resistant *Acinetobacter baumannii* by ampicillin-sulbactam and meropenem combination therapy. International Journal of Infectious Diseases. 2013;**17**(12):e1234-e1236. DOI: 10.1016/j.ijid.2013.05.002

[58] Jaruratanasirikul S, Nitchot W, Wongpoowarak W, Samaeng M, Nawakitrangsan M. Population pharmacokinetics and Monte Carlo simulations of sulbactam to optimize dosage regimens in patients with ventilator-associated pneumonia caused by *Acinetobacter baumannii*. European Journal of Pharmaceutical Sciences. 2019;**136**:104940. DOI: 10.1016/j.ejps.2019.05.018

[59] IDSA, Tamma PD, Aitken SL, Bonomo RA, Mathers AJ, van Duin D, et al. Infectious Diseases Society of America antimicrobial-resistant treatment guidance: Gram-negative bacterial infections. Infectious Diseases Society of America. 2023. Version 3.0. Available from: https://www.idsociety.org/practice-guideline/amr-guidance/ [Accessed: August 29, 2023]

[60] Paul M, Carrara E, Retamar P, Tängdén T, Bitterman R, Bonomo RA, et al. European Society of Clinical Microbiology and Infectious Diseases (ESCMID) guidelines for the treatment of infections caused by multidrug-resistant gram-negative bacilli (endorsed by European society of intensive care medicine). Clinical Microbiology and Infection. 2022;**28**(4):521-547. DOI: 10.1016/j.cmi.2021.11.025

[61] Zalts R, Neuberger A, Hussein K, Raz-Pasteur A, Geffen Y, Mashiach T, et al. Treatment of Carbapenem-resistant *Acinetobacter baumannii* ventilator-associated pneumonia: Retrospective comparison between intravenous Colistin and intravenous ampicillin-Sulbactam. American Journal of Therapeutics. 2016;**23**(1):e78-e85. DOI: 10.1097/MJT.0b013e3182a32df3

[62] Paul M, Bishara J, Levcovich A, Chowers M, Goldberg E, Singer P, et al. Effectiveness and safety of colistin: Prospective comparative cohort study. The Journal of Antimicrobial Chemotherapy. 2010;**65**(5):1019-1027. DOI: 10.1093/jac/dkq069

[63] Betrosian AP, Frantzeskaki F, Xanthaki A, Georgiadis G. High-dose ampicillin-sulbactam as an alternative treatment of late-onset VAP from multidrug-resistant *Acinetobacter baumannii*. Scandinavian Journal of Infectious Diseases. 2007;**39**(1):38-43. DOI: 10.1080/00365540600951184

[64] Kengkla K, Kongpakwattana K, Saokaew S, Apisarnthanarak A, Chaiyakunapruk N. Comparative efficacy and safety of treatment options for MDR and XDR *Acinetobacter baumannii* infections: A systematic review and network meta-analysis. The Journal of Antimicrobial Chemotherapy. 2018;**73**(1):22-32. DOI: 10.1093/jac/dkx368

[65] Krizova L, Poirel L, Nordmann P, Nemec A. TEM-1 β-lactamase as a source of resistance to sulbactam in clinical strains of *Acinetobacter baumannii*. The Journal of Antimicrobial Chemotherapy. 2013;**68**(12):2786-2791. DOI: 10.1093/jac/dkt275

[66] Yang Y, Fu Y, Lan P, Xu Q, Jiang Y, Chen Y, et al. Molecular epidemiology and mechanism of Sulbactam resistance in *Acinetobacter baumannii* isolates with diverse genetic backgrounds in China. Antimicrobial Agents and Chemotherapy. 2018;**62**(3):e01947-e01917. DOI: 10.1128/AAC.01947-17

[67] Álvarez-Marín R, López-Rojas R, Márquez JA, Gómez MJ, Molina J, Cisneros JM, et al. Colistin dosage without loading dose is efficacious when treating Carbapenem-resistant *Acinetobacter baumannii* ventilator-associated pneumonia caused by strains with high susceptibility to Colistin.

PLoS One. 2016;**11**(12):e0168468. DOI: 10.1371/journal.pone.0168468

[68] Katip W, Uitrakul S, Oberdorfer P. Clinical outcomes and nephrotoxicity of colistin loading dose for treatment of extensively drug-resistant *Acinetobacter baumannii* in cancer patients. Infection and Drug Resistance. 2017;**10**:293-298. DOI: 10.2147/IDR.S144314

[69] Falagas ME, Bliziotis IA, Tam VH. Intraventricular or intrathecal use of polymyxins in patients with gram-negative meningitis: A systematic review of the available evidence. International Journal of Antimicrobial Agents. 2007;**29**(1):9-25. DOI: 10.1016/j.ijantimicag.2006.08.024

[70] Vasen W, Desmery P, Ilutovich S, Mitre IS, Mitre B, Di Martino A. Intrathecal use of Colistin. Journal of Clinical Microbiology. 2000;**38**(9):3523. DOI: 10.1128/JCM.38.9.3523-3523.2000

[71] Jung SY, Lee SH, Lee SY, Yang S, Noh H, Chung EK, et al. Antimicrobials for the treatment of drug-resistant *Acinetobacter baumannii* pneumonia in critically ill patients: A systemic review and Bayesian network meta-analysis. Critical Care. 2017;**21**(1):319. DOI: 10.1186/s13054-017-1916-6

[72] Ye JJ, Lin HS, Yeh CF, Wu YM, Huang PY, Yang CC, et al. Tigecycline-based versus sulbactam-based treatment for pneumonia involving multidrug-resistant *Acinetobacter calcoaceticus-Acinetobacter baumannii* complex. BMC Infectious Diseases. 2016;**16**:374. DOI: 10.1186/s12879-016-1717-6

[73] Liang CA, Lin YC, Lu PL, Chen HC, Chang HL, Sheu CC. Antibiotic strategies and clinical outcomes in critically ill patients with pneumonia caused by carbapenem-resistant *Acinetobacter baumannii*. Clinical Microbiology and Infection. 2018;**24**(8):908.e1-908.e7. DOI: 10.1016/j.cmi.2017.10.033

[74] Liou BH, Lee YT, Kuo SC, Liu PY, Fung CP. Efficacy of tigecycline for secondary Acinetobacter bacteremia and factors associated with treatment failure. Antimicrobial Agents and Chemotherapy. 2015;**59**(6):3637-3640. DOI: 10.1128/AAC.04987-14

[75] De Pascale G, Montini L, Pennisi M, Bernini V, Maviglia R, Bello G, et al. High dose tigecycline in critically ill patients with severe infections due to multidrug-resistant bacteria. Critical Care. 2014;**18**(3):R90. DOI: 10.1186/cc13858

[76] Zha L, Pan L, Guo J, French N, Villanueva EV, Tefsen B. Effectiveness and safety of high dose Tigecycline for the treatment of severe infections: A systematic review and meta-analysis. Advances in Therapy. 2020;**37**(3):1049-1064. DOI: 10.1007/s12325-020-01235-y

[77] Lodise TP, Van Wart S, Sund ZM, Bressler AM, Khan A, Makley AT, et al. Pharmacokinetic and Pharmacodynamic profiling of minocycline for injection following a single infusion in critically ill adults in a phase IV open-label multicenter study (ACUMIN). Antimicrobial Agents and Chemotherapy. 2021;**65**(3):e01809-e01820. DOI: 10.1128/AAC.01809-20

[78] Boyd N, Nailor MD. Combination antibiotic therapy for empiric and definitive treatment of gram-negative infections: Insights from the society of infectious diseases pharmacists. Pharmacotherapy: The Journal of Human Pharmacology and Drug Therapy. 2011;**31**(11):1073-1084. DOI: 10.1592/phco.31.11.1073

[79] Shields RK, Paterson DL, Tamma PD. Navigating available treatment options for Carbapenem-resistant *Acinetobacter*

baumannii-calcoaceticus complex infections. Clinical Infectious Diseases. 2023;**76**(Suppl 2):S179-S193. Available from: https://pubmed.ncbi.nlm.nih.gov/37125467/. DOI: 10.1093/cid/ciad094

[80] Oleksiuk LM, Nguyen MH, Press EG, Updike CL, O'Hara JA, Doi Y, et al. In vitro responses of *Acinetobacter baumannii* to two- and three-drug combinations following exposure to Colistin and Doripenem. Antimicrobial Agents and Chemotherapy. 2014;**58**(2):1195-1199. DOI: 10.1128/AAC.01779-13

[81] Paul M, Daikos GL, Durante-Mangoni E, Yahav D, Carmeli Y, Benattar YD, et al. Colistin alone versus colistin plus meropenem for treatment of severe infections caused by carbapenem-resistant gram-negative bacteria: An open-label, randomised controlled trial. The Lancet Infectious Diseases. 2018;**18**(4):391-400. DOI: 10.1016/S1473-3099(18)30099-9

[82] Kaye KS, Marchaim D, Thamlikitkul V, Carmeli Y, Chiu CH, Daikos G, et al. Colistin monotherapy versus combination therapy for carbapenem-resistant organisms. NEJM Evidence. Jan 2023;**2**(1):10.1056/evidoa2200131. DOI: 10.1056/evidoa2200131

[83] Qureshi ZA, Hittle LE, O'Hara JA, Rivera JI, Syed A, Shields RK, et al. Colistin-resistant *Acinetobacter baumannii*: Beyond Carbapenem resistance. Clinical Infectious Diseases. 2015;**60**(9):1295-1303. DOI: 10.1093/cid/civ048

[84] Heil EL, Claeys KC, Kline EG, Rogers TM, Squires KM, Iovleva A, et al. Early initiation of three-drug combinations for the treatment of carbapenem-resistant a. baumannii among COVID-19 patients. Journal of Antimicrobial Chemotherapy. 2023;**78**(4):1034-1040. DOI: 10.1093/jac/dkad042

[85] Bremmer DN, Bauer KA, Pouch SM, Thomas K, Smith D, Goff DA, et al. Correlation of checkerboard synergy testing with time-kill analysis and clinical outcomes of extensively drug-resistant *Acinetobacter baumannii* respiratory infections. Antimicrobial Agents and Chemotherapy. 2016;**60**(11):6892-6895. DOI: 10.1128/AAC.00981-16

[86] Bonapace CR, White RL, Friedrich LV, Bosso JA. Evaluation of antibiotic synergy against *Acinetobacter baumannii*: A comparison with Etest, time-kill, and checkerboard methods. Diagnostic Microbiology and Infectious Disease. 2000;**38**(1):43-50. DOI: 10.1016/s0732-8893(00)00163-2

[87] Parchem NL, Bauer KA, Cook CH, Mangino JE, Jones CD, Porter K, et al. Colistin combination therapy improves microbiologic cure in critically ill patients with multi-drug resistant gram-negative pneumonia. European Journal of Clinical Microbiology & Infectious Diseases. 2016;**35**(9):1433-1439. DOI: 10.1007/s10096-016-2681-1

[88] Sodeifian F, Zangiabadian M, Arabpour E, Kian N, Yazarlou F, Goudarzi M, et al. Tigecycline-containing regimens and multi drug-resistant *Acinetobacter baumannii*: A systematic review and meta-analysis. Microbial Drug Resistance. 2023;**29**(8):344-359. DOI: 10.1089/mdr.2022.0248

[89] Sirijatuphat R, Thamlikitkul V. Preliminary study of Colistin versus Colistin plus Fosfomycin for treatment of Carbapenem-resistant *Acinetobacter baumannii* infections. Antimicrobial Agents and Chemotherapy. 2014;**58**(9):5598-5601. DOI: 10.1128/AAC.02435-13

[90] Durante-Mangoni E, Signoriello G, Andini R, Mattei A, De Cristoforo M, Murino P, et al. Colistin and rifampicin compared with colistin alone for the treatment of serious infections due to extensively drug-resistant *Acinetobacter baumannii*: A multicenter, randomized clinical trial. Clinical Infectious Diseases. 2013;**57**(3):349-358. DOI: 10.1093/cid/cit253

[91] Batoni G, Maisetta G, Esin S. Antimicrobial peptides and their interaction with biofilms of medically relevant bacteria. Biochimica et Biophysica Acta (BBA)—Biomembranes. 2016;**1858**(5):1044-1060. DOI: 10.1016/j.bbamem.2015.10.013

[92] Ozbek B, Mataraci E. In vitro effectiveness of colistin, tigecycline and levofloxacin alone and combined with clarithromycin and/or heparin as lock solutions against embedded *Acinetobacter baumannii* strains. The Journal of Antimicrobial Chemotherapy. 2013;**68**(4):827-830. DOI: 10.1093/jac/dks472

[93] Zavascki AP, Goldani LZ, Li J, Nation RL. Polymyxin B for the treatment of multidrug-resistant pathogens: A critical review. The Journal of Antimicrobial Chemotherapy. 2007;**60**(6):1206-1215. DOI: 10.1093/jac/dkm357

[94] Rigatto MH, Vieira FJ, Antochevis LC, Behle TF, Lopes NT, Zavascki AP. Polymyxin B in combination with antimicrobials lacking In vitro activity versus Polymyxin B in monotherapy in critically ill patients with *Acinetobacter baumannii* or *Pseudomonas aeruginosa* infections. Antimicrobial Agents and Chemotherapy. 2015;**59**(10):6575-6580. DOI: 10.1128/AAC.00494-15

[95] Liu J, Shu Y, Zhu F, Feng B, Zhang Z, Liu L, et al. Comparative efficacy and safety of combination therapy with high-dose sulbactam or colistin with additional antibacterial agents for multiple drug-resistant and extensively drug-resistant *Acinetobacter baumannii* infections: A systematic review and network meta-analysis. Journal of Global Antimicrobial Resistance. 2021;**24**:136-147. DOI: 10.1016/j.jgar.2020.08.021

[96] Mosaed R, Haghighi M, Kouchak M, Miri MM, Salarian S, Shojaei S, et al. Interim study: Comparison of safety and efficacy of levofloxacin plus Colistin regimen with levofloxacin plus high dose ampicillin/Sulbactam infusion In treatment of ventilator-associated pneumonia due to multi drug resistant Acinetobacter. Iranian Journal of Pharmaceutical Research: IJPR. 2018;**17**(Suppl 2):206-213. PMID: 31011353

[97] Abdul-Mutakabbir JC, Nguyen L, Maassen P, et al. In-vitro antibacterial activities of cefiderocol (S649266) alone and with the addition of beta-lactamase inhibitors against multidrug-resistant *Acinetobacter baumannii* (abstract 1297). Open Forum Infectious Diseases. 2020;**2**:S663. DOI: 10.1093/ofid/ofaa439.1480

[98] Garnacho-Montero J, Amaya-Villar R, Gutiérrez-Pizarraya A, Espejo-Gutiérrez de Tena E, Artero-González ML, Corcia-Palomo Y, et al. Clinical efficacy and safety of the combination of colistin plus vancomycin for the treatment of severe infections caused by carbapenem-resistant *Acinetobacter baumannii*. Chemotherapy. 2013;**59**(3):225-231. DOI: 10.1159/000356004

[99] Petrosillo N, Giannella M, Antonelli M, Antonini M, Barsic B, Belancic L, et al. Clinical experience of colistin-glycopeptide combination in critically ill patients infected with gram-negative bacteria. Antimicrobial Agents

and Chemotherapy. 2014;**58**(2):851-858. DOI: 10.1128/AAC.00871-13

[100] Lu Q, Luo R, Bodin L, Yang J, Zahr N, Aubry A, et al. Efficacy of high-dose nebulized Colistin in ventilator-associated pneumonia caused by multidrug-resistant *Pseudomonas aeruginosa* and *Acinetobacter baumannii*. Anesthesiology. 2012;**117**(6):1335-1347. DOI: 10.1097/ALN.0b013e31827515de

[101] Trottier V, Namias N, Pust DG, Nuwayhid Z, Manning R, Marttos AC, et al. Outcomes of *Acinetobacter baumannii* infection in critically ill surgical patients. Surgical Infections. 2007;**8**(4):437-443. DOI: 10.1089/sur.2006.029

[102] Magnotti LJ, Croce MA, Zarzaur BL, Swanson JM, Wood GC, Weinberg JA, et al. Causative pathogen dictates optimal duration of antimicrobial therapy for ventilator-associated pneumonia in trauma patients. Journal of the American College of Surgeons. 2011;**212**(4):476-484; discussion 484-486. DOI: 10.1016/j.jamcollsurg.2010.12.024

[103] Rodrigues RD, Garcia RCL, Bittencourt GA, Waichel VB, Garcia ECL, Rigatto MH. Antimicrobial therapy duration for bloodstream infections caused by *Pseudomonas aeruginosa* or *Acinetobacter baumannii-calcoaceticus* complex: A retrospective cohort study. Antibiotics. 2023;**12**(3):538. DOI: 10.3390/antibiotics12030538

[104] Singh N, Rogers P, Atwood CW, Wagener MM, Yu VL. Short-course empiric antibiotic therapy for patients with pulmonary infiltrates in the intensive care unit. A proposed solution for indiscriminate antibiotic prescription. American Journal of Respiratory and Critical Care Medicine. 2000;**162**(2 Pt 1):505-511. DOI: 10.1164/ajrccm.162.2.9909095

[105] Dimopoulos G, Poulakou G, Pneumatikos IA, Armaganidis A, Kollef MH, Matthaiou DK. Short- vs long-duration antibiotic regimens for ventilator-associated pneumonia: A systematic review and meta-analysis. Chest. 2013;**144**(6):1759-1767. DOI: 10.1378/chest.13-0076

[106] Chastre J, Wolff M, Fagon JY, Chevret S, Thomas F, Wermert D, et al. Comparison of 8 vs 15 days of antibiotic therapy for ventilator-associated pneumonia in adults: A randomized trial. Journal of the American Medical Association. 2003;**290**(19):2588-2598. DOI: 10.1001/jama.290.19.2588

[107] Dennesen PJ, van der Ven AJ, Kessels AG, Ramsay G, Bonten MJ. Resolution of infectious parameters after antimicrobial therapy in patients with ventilator-associated pneumonia. American Journal of Respiratory and Critical Care Medicine. 2001;**163**(6):1371-1375. DOI: 10.1164/ajrccm.163.6.2007020

[108] Tunkel AR, Hasbun R, Bhimraj A, Byers K, Kaplan SL, Scheld WM, et al. 2017 infectious diseases society of America's clinical practice guidelines for healthcare-associated ventriculitis and meningitis*. Clinical Infectious Diseases. 2017;**64**(6):e34-e65. DOI: 10.1093/cid/ciw861

[109] Falcone M, Tiseo G, Leonildi A, Della Sala L, Vecchione A, Barnini S, et al. Cefiderocol- compared to Colistin-based regimens for the treatment of severe infections caused by Carbapenem-resistant *Acinetobacter baumannii*. Antimicrobial Agents and Chemotherapy. 2022;**66**(5):e0214221. DOI: 10.1128/aac.02142-21

[110] Mazzitelli M, Gregori D, Sasset L, Trevenzoli M, Scaglione V, Lo Menzo S, et al. Cefiderocol-based versus

Colistin-based regimens for severe Carbapenem-resistant *Acinetobacter baumannii* infections: A propensity score-weighted, retrospective cohort study during the first two years of the COVID-19 pandemic. Microorganisms. 2023;**11**(4):984. DOI: 10.3390/microorganisms11040984

[111] Falcone M, Tiseo G, Nicastro M, Leonildi A, Vecchione A, Casella C, et al. Cefiderocol as rescue therapy for *Acinetobacter baumannii* and other Carbapenem-resistant gram-negative infections in intensive care unit patients. Clinical Infectious Diseases. 2021;**72**(11):2021-2024. DOI: 10.1093/cid/ciaa1410

[112] Bassetti M, Echols R, Matsunaga Y, Ariyasu M, Doi Y, Ferrer R, et al. Efficacy and safety of cefiderocol or best available therapy for the treatment of serious infections caused by carbapenem-resistant gram-negative bacteria (CREDIBLE-CR): A randomised, open-label, multicentre, pathogen-focused, descriptive, phase 3 trial. The Lancet Infectious Diseases. 2021;**21**(2):226-240. DOI: 10.1016/S1473-3099(20)30796-9

[113] Research C for DE and FDA Approves New Treatment for Pneumonia Caused by Certain Difficult-to-Treat Bacteria [Internet]. FDA. 2023. Available from: https://www.fda.gov/news-events/press-announcements/fda-approves-new-treatment-pneumonia-caused-certain-difficult-treat-bacteria [Accessed: August 24, 2023]

[114] Altarac D. Efficacy and Safety of Sulbactam Durlobactam (SUL-DUR) Versus Colistin Therpay in Patients with *Acinetobacter baumannii-Calcoaceticus* Complex (ABC) Infections: A Global, Randomised, Active-Controlled Phase 3 Trial (ATTACK). Lisbon, Portugal; 2022. DOI: 10.1016/S1473-3099(23)00184-6

[115] Seifert H, Stefanik D, Sutcliffe JA, Higgins PG. In-vitro activity of the novel fluorocycline eravacycline against carbapenem non-susceptible *Acinetobacter baumannii*. International Journal of Antimicrobial Agents. 2018;**51**(1):62-64. DOI: 10.1016/j.ijantimicag.2017.06.022

[116] Solomkin JS, Gardovskis J, Lawrence K, Montravers P, Sway A, Evans D, et al. IGNITE4: Results of a phase 3, randomized, multicenter, prospective trial of Eravacycline vs Meropenem in the treatment of complicated Intraabdominal infections. Clinical Infectious Diseases. 2019;**69**(6):921-929. DOI: 10.1093/cid/ciy1029

[117] Solomkin J, Evans D, Slepavicius A, Lee P, Marsh A, Tsai L, et al. Assessing the efficacy and safety of Eravacycline vs Ertapenem in complicated intra-abdominal infections in the investigating gram-negative infections treated with Eravacycline (IGNITE 1) trial: A randomized clinical trial. JAMA Surgery. 2017;**152**(3):224-232. DOI: 10.1001/jamasurg.2016.4237

[118] Tetraphase Pharmaceuticals. Efficacy and safety of eravacycline compared with levofloxacin in complicated urinary tract infections. Accession No. NCT01978938. ClinicalTrials.gov, NIH. 2019. Available from: https:// clinicaltrials.gov/ct2/NCT01978938

[119] Abbey T, Vialichka A, Jurkovic M, Biagi M, Wenzler E. Activity of Omadacycline alone and in combination against Carbapenem-nonsusceptible *Acinetobacter baumannii* with varying minocycline susceptibility. Microbiology Spectrum. 2022;**10**(3):e0054222. DOI: 10.1128/spectrum.00542-22

[120] Morrisette T, Alosaimy S, Lagnf AM, Frens JJ, Webb AJ, Veve MP, et al. Real-world, multicenter case series of patients treated with Oral Omadacycline for resistant gram-negative pathogens. Infectious Disease and Therapy. 2022;**11**(4):1715-1723. DOI: 10.1007/s40121-022-00645-5

[121] García-Salguero C, Rodríguez-Avial I, Picazo JJ, Culebras E. Can Plazomicin alone or in combination Be a therapeutic option against Carbapenem-resistant *Acinetobacter baumannii*? Antimicrobial Agents and Chemotherapy. 2015;**59**(10):5959-5966. DOI: 10.1128/AAC.00873-15

[122] Alfieri A, Di Franco S, Donatiello V, Maffei V, Fittipaldi C, Fiore M, et al. Plazomicin against multidrug-resistant bacteria: A scoping review. Life. 2022;**12**(12):1949. DOI: 10.3390/life12121949

[123] Chang RYK, Nang SC, Chan HK, Li J. Novel antimicrobial agents for combating antibiotic-resistant bacteria. Advanced Drug Delivery Reviews. 2022;**187**:114378. DOI: 10.1016/j.addr.2022.114378

[124] Hua Y, Luo T, Yang Y, Dong D, Wang R, Wang Y, et al. Phage therapy as a promising new treatment for lung infection caused by Carbapenem-resistant *Acinetobacter baumannii* in mice. Frontiers in Microbiology. 2017;**8**:2659. DOI: 10.3389/fmicb.2017.02659

[125] Jeon J, Park JH, Yong D. Efficacy of bacteriophage treatment against carbapenem-resistant *Acinetobacter baumannii* in galleria mellonella larvae and a mouse model of acute pneumonia. BMC Microbiology. 2019;**19**(1):70. DOI: 10.1186/s12866-019-1443-5

[126] Li Y, Xiao S, Huang G. *Acinetobacter baumannii* bacteriophage: Progress in isolation, genome sequencing, preclinical research, and clinical application. Current Microbiology. 2023;**80**(6):199. DOI: 10.1007/s00284-023-03295-z

[127] Schooley RT, Biswas B, Gill JJ, Hernandez-Morales A, Lancaster J, Lessor L, et al. Development and use of personalized bacteriophage-based therapeutic cocktails to treat a patient with a disseminated resistant *Acinetobacter baumannii* infection. Antimicrobial Agents and Chemotherapy. 2017;**61**(10):e00954-e00917. DOI: 10.1128/AAC.00954-17

[128] Choi J, Jang A, Yoon YK, Kim Y. Development of novel peptides for the antimicrobial combination therapy against Carbapenem-resistant *Acinetobacter baumannii* infection. Pharmaceutics. 2021;**13**(11):1800. DOI: 10.3390/pharmaceutics13111800

Chapter 5

Current Options for the Treatment of Urinary Tract Infections Caused by Multiresistant *Acinetobacter baumannii*

Alexandre Bueno Merlini, Renan Henrique Merlini, Bernardo Noya de Abreu and Maria Daniela Di Dea Bergamasco

Abstract

Urinary tract infections (UTIs) are the main etiological agent of Gram-negative bacteria. UTI and pneumonia are the main causes of sepsis in older people. With the advance of medicine, the increase in life expectancy, more frequent prescription of immunosuppressive therapies, and indiscriminate use of antibiotics, multidrug-resistant (MDR) pathogens have become a global public health problem. Among them, the rise of MDR *Acinetobacter baumannii* infections is observed in hospitals, especially in patients accommodated in intensive care units (ICU) and/or in the use of medical devices, such as urinary catheters. Treating UTIs caused by carbapenem-resistant *Acinetobacter baumannii* became a challenge, given the few therapeutic options and low penetration of polymyxin B into the renal parenchyma.

Keywords: urinary tract infection, *Acinetobacter baumannii*, multidrug-resistant, antibiotics, treatment

1. Introduction

Urinary tract infection (UTI) is one of the most common infections in humans. The Gram-negative bacteria represent the main etiological group in community and nosocomial cases [1, 2].

Nowadays, *Acinetobacter baumannii* — a coccobacillus Gram-negative — is an important pathogen to hospitals worldwide, becoming a public health problem when multidrug-resistant (MDR) [3]. Its ability to overlap resistance mechanisms culminated in the appearance of strains resistant to all available antibiotics in the industry [4]. Most infections are healthcare-associated and linked to invasive devices such as urinary catheters [5].

The World Health Organization (WHO) listed carbapenem-resistant strains as one of the priority agents for developing new antibiotics [6]. This chapter aims to bring options for treating UTI caused by MDR *A. baumannii*, given that the urinary concentration of drugs restricts the choice of the therapeutic regimen [7, 8].

Risk factors
ICU hospitalization (previous or current)
Recent surgical procedures
Previous colonization by Methicillin-Resistant *Staphylococcus aureus* (MRSA)
Invasive devices such as central venous and urinary catheters
Hemodialysis
Malignant neoplasms
Previous administration of beta-lactams (mainly carbapenems) and fluoroquinolones
Infusion of neurobiological or chemotherapeutic agents
Bed restriction
Burns
Preterm birth

Table 1.
Risk factors involving infections caused by MDR A. baumannii.

2. Epidemiology

MDR strains have become frequent causes of nosocomial infections since the 1980s. In a study from 2007 that involved 100 hospitals worldwide, 34% of *Acinetobacter* isolates were resistant to ceftazidime and 41% to ciprofloxacin. From 1999 to that year, the carbapenem resistance increased from 10 to 54% [9].

In a recent study conducted by Seifer et al., between 2016 and 2018, the global resistance to meropenem reached 67%, while the overall resistance to colistin was 7%. The highest percentage of carbapenem resistance of over 90% was reported in the Mediterranean region, imposing serious burdens on healthcare systems [10].

Di Venanzio et al. analyzed *Acinetobacter* isolates identified in the BJC Healthcare System from January 2007 to August 2017. The study showed that, among the over 19,000 cases, 17.1% came from the urinary tract [11]. But only 2% of UTIs are caused by this pathogen. However, *A. baumannii* is the main agent causing UTIs associated with using catheters in ICUs. More than 50% of the isolated strains from urine come from catheterized patients [12].

There are some risk factors for developing infections caused by MDR *A. baumannii* (**Table 1**) [13–16].

3. Clinical presentation

The signs and symptoms vary according to the affected segment of the urinary tract. The main manifestations of acute cystitis include dysuria, pollakiuria, suprapubic pain, urinary urgency, and even hematuria. In older patients, the identification could be more difficult due to a higher frequency of nonspecific symptoms. Those patients can present delirium, change in level of consciousness, prostration, and inappetence [17].

Fever and other systemic symptoms (nausea, vomiting, and nonmechanical back pain) suggest a complicated UTI or upper urinary tract involvement. Hypotension, tachycardia, tachypnea, and oliguria suggest a more severe infection, such as sepsis and septic shock [18].

The anamnesis may assess, beyond the medical history, the use (current or prior) of invasive devices, particularly urinary catheterization, recent hospitalization (especially in ICU and emergency departments), and recent use of antibiotics [19].

It is fundamental to differentiate a context of infection from colonization, which will lead to different approaches [20, 21].

4. Diagnosis

In addition to the clinical evaluation, it is recommended to request complementary exams. In front of a suspicion of a not complicated UTI, it is important to perform a urinalysis and urine culture. Patients with preserved consciousness and urinary continence may spontaneously collect a midstream urine specimen after proper hygiene of the genitourinary region [22].

In patients with systemic symptoms, especially the elderly, diabetic, with an immunosuppressive condition, blood cultures and imaging exams (ultrasound or computed tomography) must be performed to screen for pyelonephritis or complications, such as kidney abscess (**Figure 1**) [23].

5. Treatment

In a first medical evaluation, in front of a urinary tract infection, the physician will not know the etiological agent, even if the patient has risk factors for MDR pathogens, and an empiric treatment would be initiated. In these cases, broad-spectrum antibiotics are recommended with coverage for Gram-negative bacilli. Prior — and particularly current — cultures can guide the chosen scheme. If the patient uses a urinary catheter, it is part of the treatment to remove or exchange it for a new one [24, 25].

After identifying a strain of *A. baumannii* in the cultures, the treatment must be based on the sensitivity profile of the antibiogram (**Figure 2**).

The therapeutic scheme is divided into first-line, second-line, and synergistic agents (not recommended monotherapy) [26].

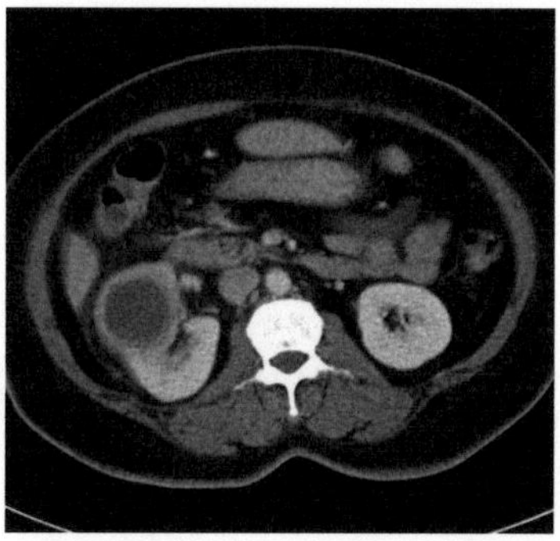

Figure 1.
Perinephric abscess compromising the mid pole of right kidney, with thickening of Gerota's fascia. Case courtesy of Ian Bickle, Radiopaedia.org, rID: 29853.

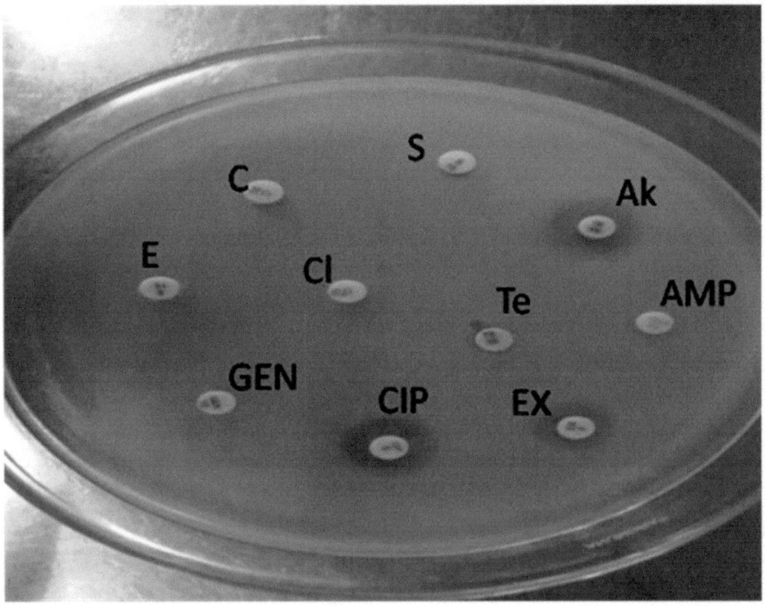

Figure 2.
An example of an antibiogram with a MDR pathogen. Ak = amikacin; AMP = ampicillin; C = chloramphenicol; Cl = colistin; CIP = ciprofloxacin; E = erythromycin; EX = enrofloxacin; GEN = gentamicin; S = streptomycin; Te = tetracycline.

5.1 First-line agents

When susceptible to the antibiogram, a first-line antibiotic must be chosen. The main options are listed below with the indicated dosage for a patient with normal renal function (**Table 2**) [27–32].

5.2 Second-line agents

When it is diagnosed as a resistant infection to all the first-line agents, one of the options below must be considered (**Table 3**) [33–36].

Antibiotic	Dosage
Ciprofloxacin	750 mg orally bid or 400 mg every 8 hours
Levofloxacin	750 mg orally or intravenously qd
Trimethoprim-sulfamethoxazole	1 tablet (160 + 800 mg) bid
Ampicillin-sulbactam	Carbapenem-susceptible infections: 3 g intravenously every 6 hours carbapenem-resistant conditions: 3 g intravenously every 4 hours Severe conditions: 9 g intravenously every 8 hours (or 27 g at continuous infusion)
Ceftazidime	2 g intravenously every 8 hours
Cefepime	2 g intravenously every 8 hours
Tazobactam-piperacillin	4.5 g intravenously every 6 or 8 hours

Antibiotic	Dosage
Meropenem	Cystitis: 1 g intravenously every 8 hours Pyelonephritis or complicated infections: 2 g intravenously every 8 hours
Imipenem	Cystitis: 500 mg intravenously every 6 hours Pyelonephritis or complicated infections: 500 mg to 1 g every 6 or 8 hours
Gentamicin	Cystitis: 5 mg/kg intravenously qd
Amikacin	Cystitis: 15 mg/kg intravenously qd

Table 2.
First-line antibiotics used in the treatment of MDR A. baumannii.

5.3 Combined therapy

Some antibiotics, even with adequate sensitivity on antibiogram, could be ineffective as monotherapy. Although aminoglycosides are a good option for treating mild cases of UTI, their isolated administration is not recommended for moderate and severe infections (**Table 4**) [32, 37, 38].

5.4 Other considerations

To treat cystitis without systemic manifestations, some experts recommend the oral administration of fosfomycin. However, its effectiveness could be better. *A. baumannii* is intrinsically resistant *in vitro* studies. No protocols define the duration of treatment, and there is no standardized methodology to determine susceptibility [39, 40].

Doxycycline, minocycline, and tigecycline usually do not present enough serum or urinary concentrations to treat UTI properly [41–43].

Antibiotic	Dosage
Colistin (polymyxin E)	Loading dosage of 9 million units of colistimethate sodium. Daily maintenance dosage of 9 to 11 million units, divided into three or three infusions.
Cefiderocol	2 g every 8 hours - this medication was approved by the Food and Drug Administration (FDA) for complicated UTIs in 2019. It is only available in some countries.

Table 3.
Second-line antibiotics used in the treatment of MDR A. baumannii.

Antibiotic	Dosage
Polymyxin B	Zavascki et al. demonstrated in a study that only 1% of the unaltered drug was found in the urine. Therefore, colistin is preferred for the treatment of UTIs. When colistin is unavailable, the recommended loading dose of polymyxin B is 20.000 units/kg, followed by a maintenance dose of 15.000 units/kg bid.
Gentamicin	Pyelonephritis or complicated infections: 7 mg/kg intravenously for the first dose, followed by 2–3 mg/kg/day divided into two or three doses.
Amikacin	Pyelonephritis or complicated infections: 20 mg/kg intravenously for the first dose, followed by 15 mg/kg/day, in a single dose or divided into two doses.

Table 4.
Antibiotics used as part of combined therapy in the treatment of MDR A. baumannii.

Novel antibiotics should not be used because they have limited *in vitro* activity against *Acinetobacter* strains [44, 45].

6. Conclusion

The incidence of MDR *Acinetobacter* has increased in recent decades with higher resistance to carbapenems and colistin. To treat these infections became a challenge and a public health problem. When analyzing the management of UTI-caused MDR *A. baumannii*, the options become scarce because of the low urinary concentration of some drugs. Novel agents until now are ineffective, owing to an observed *in vitro* intrinsic resistance. Therefore, developing new antibiotics, and even vaccines, is necessary and is in the sights of scholars as part of the WHO's goals for the near future.

Conflict of interest

The authors declare no conflict of interest.

Appendices and nomenclature

bid	two times a day;
qd	once time a day;
ICU	intensive care unit
MDR	multidrug-resistant
MRSA	methicillin-resistant *Staphylococcus aureus*
UTI	urinary tract infection
WHO	world health organization

Author details

Alexandre Bueno Merlini[1*], Renan Henrique Merlini[2], Bernardo Noya de Abreu[1] and Maria Daniela Di Dea Bergamasco[3]

1 Department of Emergency Medicine, Hcor, São Paulo, Brazil

2 Department of Medicine, Universidade Nove de Julho, São Paulo, Brazil

3 Department of Infectious Diseases, Hcor, São Paulo, Brazil

*Address all correspondence to: alexandre.merlini@gmail.com

References

[1] Paul M, Carrara E, Retamar P, Tängdén T, Bitterman R, Bonomo RA, et al. European society of clinical microbiology and infectious diseases (ESCMID) guidelines for treating infections caused by multidrug-resistant gram-negative bacilli (endorsed by the European society of intensive care medicine). Clinical Microbiology and Infection. 2022;**28**(4):521-547. DOI: 10.1016/j.cmi.2021.11.025

[2] Jiménez-Guerra G, Heras-Cañas V, Gutiérrez-Soto M, Aznarte-Padial MDP, Expósito-Ruiz M, Navarro-Marí JM, et al. Urinary tract infection by *Acinetobacter baumannii* and *Pseudomonas aeruginosa*: Evolution of antibiotic resistance and therapeutic alternatives. Journal of Medical Microbiology. 2018;**67**(6):790-797. DOI: 10.1099/jmm.0.000742

[3] Jung SY, Lee SH, Lee SY, Yang S, Noh H, Chung EK, et al. Antimicrobials for treating drug-resistant *Acinetobacter baumannii* pneumonia in critically ill patients: A systemic review and Bayesian network meta-analysis. Critical Care. 2017;**21**(1):319. DOI: 10.1186/s13054-017-1916-6

[4] Higgins PG, Dammhayn C, Hackel M, Seifert H. Global spread of carbapenem-resistant *Acinetobacter baumannii*. The Journal of Antimicrobial Chemotherapy. 2010;**65**(2):233-238. DOI: 10.1093/jac/dkp428

[5] Papanikolopoulou A, Maltezou MC, Stoupis A, Kalimeri D, Pavli A, Bonfidou F, et al. Catheter-associated urinary tract infections, bacteremia, and infection control interventions in a hospital: A six-year time-series study. Journal of Clinical Medicine. 2022;**11**(18):5418. DOI: 10.3390/jcm11185418

[6] Tacconelli E, Carrara E, Savoldi A, Harbarth S, Mendelson M, Monnet DL, et al. Discovery, research, and development of new antibiotics: The WHO priority list of antibiotic-resistant bacteria and tuberculosis. The Lancet Infectious Diseases. 2018;**18**(3):318. DOI: 10.1016/S1473-3099(17)30753-3

[7] Qureshi ZA, Hittle LE, O'Hara JA, Rivera JI, Syed A, Shields RK, et al. Colistin-resistant *Acinetobacter baumannii*: Beyond carbapenem resistance. Clinical Infectious Diseases. 2015;**60**(9):1295-1303. DOI: 10.1093/cid/civ048

[8] Kotov SV, Pulbere SA, Alesina NV, Boyarkin VS, Guspanov RI, Belomystev SV, et al. The problem of antibiotic resistance in patients with urinary tract infection. Urologiia. 2021;**1**:5-12

[9] Sunenshine RH, Wright MO, Maragakis LL, Harris AD, Song X, Hebden J, et al. Multidrug-resistant *Acinetobacter* infection mortality rate and length of hospitalization. Emerging Infectious Diseases. 2007;**13**(1):97. DOI: 10.3201/eid1301.060716

[10] Seifert H, Blondeau J, Lucaßen K, Utt EA. Global update on the in vitro activity of tigecycline and comparators against isolates of *Acinetobacter baumannii* and rates of resistant phenotypes (2016-2018). Journal of Global Antimicrobial Resistance. 2022;**31**:82-89. DOI: 10.1016/j.jgar.2022.08.002

[11] Di Venanzio G, Flores-Mireles AL, Calix JJ, Haurat MF, Scott NE, Palmer LD, et al. Urinary tract colonization is enhanced by a plasmid that regulates uropathogenic *Acinetobacter*

baumannii chromosomal genes. Nature Communications. 2019;**10**(1):2763. DOI: 10.1038/s41467-019-10706-y

[12] Mohamed AH, Omar NMS, Osman MM, Mohamud HA, Eraslan A, Gur M. Antimicrobial resistance and predisposing factors associated with catheter-associated UTI caused by uropathogenic exhibiting multidrug-resistant patterns: A three year-retrospective study at a tertiary hospital in Mogadishu, Somalia. Tropical Medicine and Infectious Disease. 2022;7(3):42. DOI: 10.3390/tropicalmed7030042

[13] Lee SO, Kim NJ, Choi SH, Hyong Kim T, Chung JW, Woo JH, et al. Risk factors for acquisition of imipenem-resistant *Acinetobacter baumannii*: A case-control study. Antimicrobial Agents and Chemotherapy. 2004;**48**(1):224-228. DOI: 10.1128/AAC.48.1.224-228.2004

[14] Vitkauskiene A, Dambrauskiene A, Cerniauskiene K, Rimdeika R, Sakalauskas R. Risk factors and outcomes in patients with carbapenem-resistant *Acinetobacter* infection. Scandinavian Journal of Infectious Diseases. 2013;**45**(3):213-218. DOI: 10.3109/00365548.2012.724178

[15] Jean SS, Hsueh PR. High burden of antimicrobial resistance in Asia. International Journal of Antimicrobial Agents. 2011;**37**(4):291-295. DOI: 10.1016/j.ijantimicag.2011.01.009

[16] Wu D, Chen C, Liu T, Jia Y, Wan Q, Peng J. Epidemiology, susceptibility, and risk factors associated with mortality in carbapenem-resistant gram-negative bacterial infections among abdominal solid organ transplant recipients: A retrospective cohort study. Infectious Disease and Therapy. 2021;**10**(1):559-573. DOI: 10.1007/s40121-021-00411-z

[17] Kanj SS, Bassetti M, Kiratisin P, Rodrigues C, Villegas MV, Yu Y, et al. Clinical data from studies involving novel antibiotics to treat multidrug-resistant gram-negative bacterial infections. International Journal of Antimicrobial Agents. 2022;**60**(3):106633. DOI: 10.1016/j.ijantimicag.2022.106633

[18] Wong D, Nielsen TB, Bonomo RA, Pantapalangkoor P, Luna B, Spellberg B. Clinical and pathophysiological overview of *Acinetobacter* infections: A century of challenges. Clinical Microbiology Reviews. 2017;**30**(1):409-447. DOI: 10.1128/CMR.00058-16

[19] Moghnieh RA, Kanafani ZA, Tabaja HZ, Sharara SL, Awad LS, Kanj SS. Epidemiology of common resistant bacterial pathogens in the countries of the Arab League. The Lancet Infectious Diseases. 2018;**18**(12):e379-e394. DOI: 10.1016/S1473-3099(18)30414-6

[20] Chuang L, Ratnayake L. Overcoming challenges of treating extensively drug-resistant *Acinetobacter* baumannii bacteremia urinary tract infection. International Journal of Antimicrobial Agents. 2018;**52**(4):521-522. DOI: 10.1016/j.ijantimicag.2018.07.016

[21] Bassetti M, Vena A, Giacobbe DR, Castaldo N. Management of infections caused by multidrug-resistant gram-negative pathogens: Recent advances and future directions. Archives of Medical Research. 2021;**52**(8):817-827. DOI: 10.1016/j.arcmed.2021.09.002

[22] Zohoun A, Ngoh E, Bajjou T, SekhsokhY, ElhamzaouiS. Epidemiological features of multidrug-resistant bacteria isolated from urine samples at the Mohammed V Military Teaching Hospital in Rabat, Morocco. Médecine Tropicale: Revue du Corps de santé Colonial. 2010;**70**(4):412-413

[23] Mamari YA, Sami H, Siddiqui K, Tahir HB, Jabri ZA, Muharrmi ZA, et al. Trends of antimicrobial resistance in patients with complicated urinary tract infection: Suggested empirical therapy and lessons learned from a retrospective observational study in Oman. Urology Annals. 2022;**14**(4):345-352. DOI: 10.4103/ua.ua_67_22

[24] Kumar S, Anwer R, Azzi A. Virulence potential and treatment options of multidrug-resistant (MDR) *Acinetobacter baumannii*. Microorganisms. 2021;**9**(10):2104. DOI: 10.3390/microorganisms9102104

[25] Alves MJ, Barreira JCM, Carvalho I, Trinta L, Perreira L, Ferreira ICFR, et al. Propensity for biofilm formation in clinical isolates from urinary tract infections: Developing a multifactorial predictive model to improve antibiotherapy. Journal of Medical Microbiology. 2014;**63**(Pt. 3):471-477. DOI: 10.1099/jmm.0.071746-0

[26] Isler B, Doi Y, Bonomo RA, Paterson DL. New treatment options against carbapenem-resistant *Acinetobacter baumannii* infections. Antimicrobial Agents and Chemotherapy. 2019;**63**(1):e01110-e01118. DOI: 10.1128/AAC.01110-18

[27] Heinemann B, Wisplinghoff H, Edmond M, Seifert H. Comparative activities of ciprofloxacin, clinafloxacin, gatifloxacin, gemifloxacin, levofloxacin, moxifloxacin, and trovafloxacin against epidemiologically defined *Acinetobacter baumannii* strains. Antimicrobial Agents and Chemotherapy. 2000;**44**(8):2211. DOI: 10.1128/AAC.44.8.2211-2213.2000

[28] Palka A, Kujawska A, Hareza DA, GadjaM,WordliczekJ,Jachowicz-MatczakJ, et al. Secondary bacterial infections and extensively drug-resistant bacteria among COVID-19 hospitalized patients in the University Hospital in Kraków. Annals of Clinical Microbiology and Antimicrobials. 2023;**22**(1):77. DOI: 10.1186/s12941-023-00625-8

[29] Falagas ME, Vardakas KZ, Roussos NS. Trimethoprim/sulfamethoxazole for *Acinetobacter* spp.: A review of current microbiological and clinical evidence. International Journal of Antimicrobial Agents. 2015;**46**(3):231-241. DOI: 10.1016/j.ijantimicag.2015.04.002

[30] Oliveira MS, Prado GV, Costa SF, Grinbaum RS, Levin AS. Ampicillin/sulbactam compared with polymyxins for the treatment of infections caused by carbapenem-resistant *Acinetobacter* spp. The Journal of Antimicrobial Chemotherapy. 2008;**61**(6):1369. DOI: 10.1093/jac/dkn128

[31] Dickstein Y, Lellouche J, Schwartz D, Nutman A, Rakovitsky N, Dishon Benattar Y, et al. Colistin resistance development following colistin-meropenem combination therapy versus colistin monotherapy in patients with infections caused by carbapenem-resistant organisms. Clinical Infectious Diseases. 2020;**71**(10):2599-2607. DOI: 10.1093/cid/ciz1146

[32] Liu JY, Wang FD, Ho MW, Lee CH, Liu JW, Wang JT, et al. In vitro activity of aminoglycosides against clinical isolates of *Acinetobacter baumannii* complex and other nonfermentative gram-negative bacilli causing healthcare-associated bloodstream infections in Taiwan. Journal of Microbiology, Immunology, and Infection. 2016;**49**(6):918-923. DOI: 10.1016/j.jmii.2015.07.010

[33] Tsuji BT, Pogue JM, Zavascki AP, Paul M, Daikos GL, Forrest A, et al. International consensus guidelines for the optimal use of the polymyxins:

Endorsed by the American College of Clinical Pharmacy (ACCP), European Society of Clinical Microbiology and Infectious Diseases (ESCMID), Infectious Diseases Society of America (IDSA), International Society for Anti-infective Pharmacology (ISAP), Society of Critical Care Medicine (SCCM), and Society of Infectious Diseases Pharmacists (SIDP). Pharmacotherapy. 2019;**39**(1):10-39. DOI: 10.1002/phar.2209

[34] Luque S, Escaño C, Sorli L, Li J, Campillo N, Horcajada JP, et al. Urinary concentrations of colistimethate formed colistin after intravenous administration in patients with multidrug-resistant gram-negative bacterial infections. Antimicrobial Agents and Chemotherapy. 2017;**61**(8):e02595-e02516. DOI: 10.1128/AAC.02595-16

[35] Kazmierczak KM, Tsuji M, Wise MG, Hackel M, Yamano Y, Echols R, et al. In vitro activity of cefiderocol, a siderophore cephalosporin, against a recent collection of clinically relevant carbapenem-non-susceptible gram-negative bacilli, including serine carbapenemase- and metallo-β-lactamase-producing isolates (SIDERO-WT-2014 study). International Journal of Antimicrobial Agents. 2019;**53**(2):177-184. DOI: 10.1016/j.ijantimicag.2018.10.007

[36] Sansone P, Giaccari LG, Coppolino F, Aurilio C, Barbarisi A, Passavanti MB, et al. Cefiderocol for carbapenem-resistant bacteria: Handle with care! A review of the real-world evidence. Antibiotics (Basel). 2022;**11**(7):904. DOI: 10.3390/antibiotics11070904

[37] Sacco F, Visca P, Runci F, Antonelli G, Raponi G. Susceptibility testing of colistin for *Acinetobacter baumannii*: How far are we from the truth? Antibiotics (Basel). 2021;**10**(1):48. DOI: 10.3390/antibiotics10010048

[38] Aslan AT, Akova M. The role of colistin in the era of new B-lactam/B-lactamase inhibitor combinations. Antibiotics (Basel). 2022;**11**(2):277. DOI: 10.3390/antibiotics11020277

[39] Sharma A, Sharma R, Bhattacharyya T, Bhando T, Pathania R. Fosfomycin resistance in *Acinetobacter baumannii* is mediated by efflux through a major facilitator superfamily (MFS) transporter-AbaF. The Journal of Antimicrobial Chemotherapy. 2017;**72**(1):68-74. DOI: 10.1093/jac/dkw382

[40] Ito R, Mustapha MM, Tomich AD, Callaghan JD, McElheny CL, Mettus RT, et al. Widespread fosfomycin resistance in gram-negative bacteria attributable to the chromosomal fosA gene. MBio. 2017;**8**(4):e00749-e00717. DOI: 10.1128/mBio.00749-17

[41] Peleg AY, Adams J, Paterson DL. Tigecycline efflux as a mechanism for nonsusceptibility in *Acinetobacter baumannii*. Antimicrobial Agents and Chemotherapy. 2007;**51**(6):2065. DOI: 10.1128/AAC.01198-06

[42] Flamm RK, Shortridge D, Castanheira M, Sader HS, Pfaller MA. In vitro activity of minocycline against U.S. isolates of *Acinetobacter baumannii-Acinetobacter calcoaceticus* species complex, *Stenotrophomonas maltophilia*, and *Burkholderia cepacia* complex: Results from the SENTRY antimicrobial surveillance program, 2014 to 2018. Antimicrobial Agents and Chemotherapy. 2019;**63**(11):e01154-e01119. DOI: 10.1128/AAC.01154-19

[43] Falagas ME, Vardakas KZ, Kapaskelis A, Triarides NA, Roussos NS.

Tetracyclines for multidrug-resistant *Acinetobacter baumannii* infections. International Journal of Antimicrobial Agents. 2015;**45**(5):455-460. DOI: 10.1016/j.ijantimicag.2014.12.031

[44] Rangel K, Chagas TPGC, De-Simone SG. *Acinetobacter baumannii* infections in times of COVID-19 pandemic. Pathogens. 2021;**10**(8):1006. DOI: 10.3390/pathogens10081006

[45] Arya R, Goldner BS, Shorr AF. Novel agents in development for multidrug-resistant gram-negative infections: Potential new options facing multiple challenges. Current Opinion in Infectious Diseases. 2022;**35**(6):589-594. DOI: 10.1097/QCO.0000000000000885

Section 3

Toxicity of Polymyxins in *Acinetobacter baumannii*

Chapter 6

Understanding the Harmful Impact of Polymyxins on *Acinetobacter baumannii*

Karyne Rangel, Thiago Pavoni Gomes Chagas and Salvatore Giovanni De-Simone

Abstract

Nosocomial infections caused by carbapenem-resistant *Acinetobacter baumannii* (CRAB) have become a global concern. The extensive antibiotic resistance of CRAB has significantly limited treatment options, while its prevalence in hospital outbreaks has amplified infection rates. This scenario has led to a resurgence of interest in polymyxins, an older class of antibiotics previously overlooked due to perceived toxicity. Polymyxins, cationic polypeptide antibiotics, now represent a last-resort treatment option. Despite their historical use, modern assessment methods have only recently been applied to evaluate polymyxins. Two polymyxins are available for clinical use: polymyxin B and colistin (polymyxin E). Notably, the administration of these drugs is hindered by toxicities, primarily nephrotoxicity and neurotoxicity, alongside less common adverse effects such as injection pain, hypersensitivity reactions, and bronchospasms.

Keywords: *Acinetobacter baumannii*, polymyxin, toxicity, nephrotoxicity, neurotoxicity

1. Introduction

Antimicrobial resistance (AMR) has escalated into a global healthcare crisis, rendering many pathogens resistant to current treatments [1]. A comprehensive analysis estimated 1.27 million deaths attributable to bacterial AMR in 2019 [2], and projections indicate that 2050 annual AMR-related deaths could reach ten million [3].

Over the past three decades, *Acinetobacter baumannii* has emerged as a formidable healthcare challenge, particularly due to multidrug-resistant (MDR) strains, resistant even to carbapenems [4, 5]. MDR rates for *A. baumannii* surpass those of other nosocomial pathogens [6]. *A. baumannii*, a Gram-negative non-fermentative coccobacillus of the *Moraxellacecae* family, thrives in healthcare settings owing to its antibiotic resistance and desiccation tolerance [7].

Managing *A. baumannii* infections is complex due to its diverse resistance mechanisms, with carbapenem resistance (CR) being particularly concerning. The World Health Organization (WHO) classifies carbapenem-resistant *A. baumannii* (CRAB) as a critical priority, given its threat to human health [8]. During the SARS-CoV-2 pandemic, CRAB infections further complicated patient outcomes, with high resistance rates (91.2%) observed [9].

A significant subset of CRAB isolates is extensively drug-resistant (XDR; i.e., non-susceptible to ≥1 agent in all but ≤2 classes) or pan drug-resistant (PDR; i.e., non-susceptible to all antimicrobial agents have been reported worldwide) [10–12], compounding the challenge. Limited effective antibiotic options against CRAB pose a substantial health challenge. Polymyxins, though previously overshadowed, regained prominence in the late 1990s due to their activity against carbapenem-resistant (CR) infections [13]. However, new-generation antimicrobials, particularly β-lactam/β-lactamase inhibitors, have largely replaced polymyxins in CR Gram-negative bacterial infections. Conversely, polymyxins are vital for tackling resistant pathogens [13–15], especially where new agents are unavailable [16]. Nonetheless, they come with adverse effects, including allergic reactions, neurotoxicity, and nephrotoxicity [17].

2. Polymyxins

2.1 History of discovery

Polymyxins are cationic polypeptide antibiotics derived from *Bacillus polymyxa*, pivotal in treating carbapenem-resistant Gram-negative bacteria. The initial anti-bacterial activity was reported in 1947 [18, 19], leading to the isolation of antibiotics named polymyxin [20] and aerosporin [18, 21]. Despite the structural similarity, they were classified as belonging to the same class [22–25]. Polymyxin B and polymyxin E (colistin) differ in a single amino acid (D-Phe replaces D-Leu) [26, 27] and are the clinical variants among over 15 known polymyxins [13–15, 28, 29]. These peptides share a cyclic ring structure with hydrophilic and hydrophobic components, enabling them to disrupt cell membranes [13, 29, 30].

2.2 Structure

Polymyxins' structure resembles antimicrobial peptides deployed by eukaryotes against pathogens. They are natural non-ribosomal cyclic lipopeptides weighing around 1.2 kDa (**Figure 1**) and consist of a cyclic ring of amino acids with a tripeptide chain, which binds to the lipid part of the molecule. The decapeptide core of poly-myxins contains an intramolecular loop of starch-linked heptapeptides between the amino group on the side chain of the aminobutyric acid (Dab) residue at position four and the carboxyl group on the C-terminal threonine residue. They also have several other distinctive structural features, including five non-proteogenic Dab residues positively charged at physiological pH, conserved hydrophobic residues at positions 6 and 7, and an N-terminal acyl group [31]. The cationic peptide ring of these antibiot-ics is the same between the two polymyxins, except for a single amino acid: a D-Leu from colistin is relocated by D-Phe to polymyxin B [14, 26, 27, 29–32]. However, the pharmacokinetics of polymyxin B and colistin differ notably due to the different pharmaceutical forms in which they are administered—active and prodrug form, respectively [33]. Its mechanisms of action occur through the rupture of the external and cytoplasmic membranes of the bacteria, causing loss of the contents of the cell's interior [34]. Polymyxin B comprises at least four components and polymyxin B1 to B4, which differ only in the portion containing fatty acids, polymyxin B1 and B2 being in greater proportion [35].

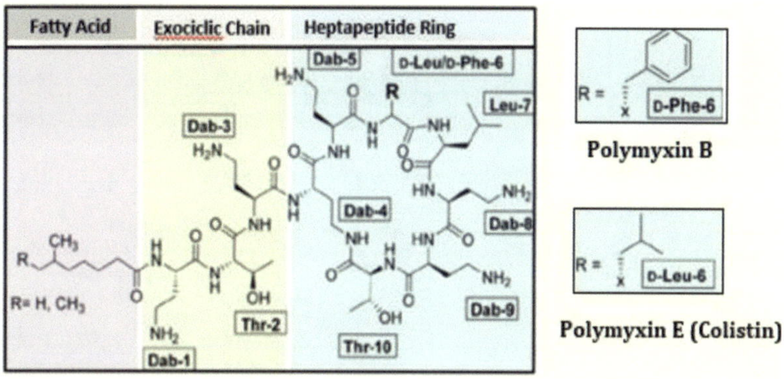

Figure 1.
Cyclic lipopeptide structure of polymyxin B (1). Colistin (polymyxin E) features a substitution of one (D-Leu) with one (D-Phe) (2).

2.3 Mechanism of action

Polymyxins exert rapid bactericidal effects by interacting with lipopolysaccharides (LPS) in the bacterial outer membrane, inducing disruptions that compromise membrane integrity. LPS, a critical component of the bacterial outer membrane, encompasses the O antigen, polysaccharide core, and lipid A. The positive charge of the polymyxin ring facilitates its binding to the outer membrane's lipid A, leading to the displacement of stabilizing Mg^2 and Ca^2 ions, which is crucial for LPS integrity [35]. The fatty acid side chains also engage with LPS, enabling the secure insertion of polymyxin into the outer membrane. This interaction triggers a series of detrimental effects, including changes in outer membrane permeability, leakage of cell contents, and eventual bacterial cell death [29, 36]. Beyond inducing cytoplasmic leakage, this binding may neutralize the biological properties of endotoxins [14, 29]. Multiple hypotheses and models exist to explain the various mechanisms underlying polymyxin's bactericidal activity [13, 14, 29]. The principal pathways through which polymyxins exhibit their activity are shown in **Figure 2**.

2.4 Polymyxin resistance

The resistance of microorganisms to polymyxin remains incompletely understood, potentially arising from mutation or adaptation mechanisms [37, 38]. In most Gram-negative bacteria, the PhoP/Q and PmrA/B regulatory systems are pivotal in mediating polymyxin resistance. These systems oversee mechanisms that induce chemical modifications in the structure of bacterial lipopolysaccharides (LPS) (**Figure 3**). In response to low levels of antimicrobial peptides, Mg+2 and Ca+2 ions, as well as other inducers such as low pH, excessive Fe+3, excessive Al+3, and phagosomes, these systems modulate resistance by altering the cationic charge of the cell wall. Cumulatively, these modifications reduce the negative charge of the bacterial outer membrane, resulting in a diminished affinity of polymyxin for the bacterial cell surface [29].

Modifying lipid A within the lipopolysaccharide (LPS) molecule, catalyzed by the gene products of pmrCAB and arnBCADTEF, is a fundamental mechanism underlying bacterial resistance to polymyxin antibiotics. These gene products play a pivotal

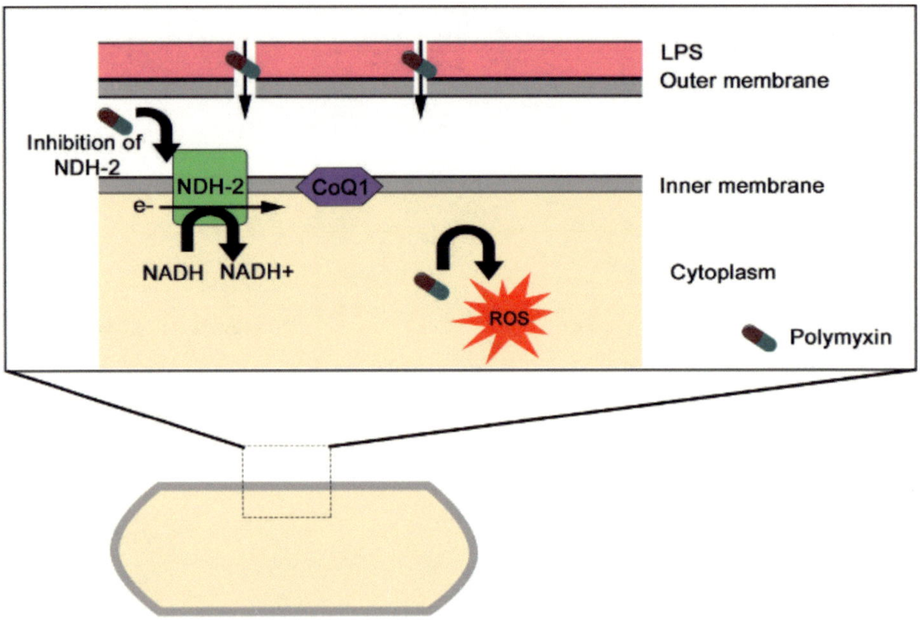

Figure 2.
Mechanisms of antibacterial activity of polymyxins in gram-negative bacteria. Disruption of the outer membrane, vesicle-vesicle contact, inhibition of respiratory enzyme NDH-2, and hydroxyl radical formation. CoQ1, coenzyme Q1.

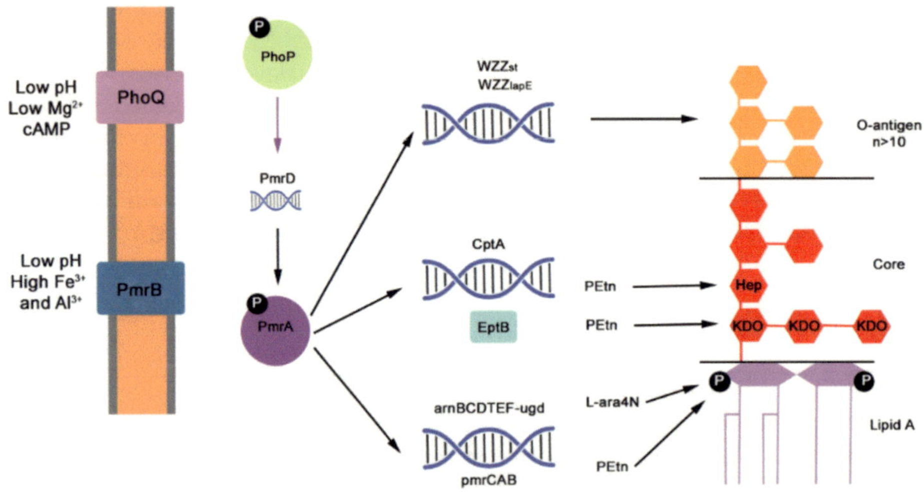

Figure 3.
Mechanism of polymyxin resistance changes in LPS. The PhoQP two-component system triggers pmrD expression. PmrD activates PmrA, cptA pmr, and the am operon. Working alongside EptB, CptA brings about modifications in the core polysaccharide of LPS. The pmr and am products facilitate the substitution of lipid a phosphates by Petn and L-ara4N, respectively. These collective alterations influence the charge of the outer membrane, resulting in polymyxin repulsion.

role in altering the surface charge and permeability of the bacterial outer membrane (OM) [39–41].

In *A. baumannii*, resistance development is primarily associated with changes in the LPS biosynthesis pathway. Currently, two mechanisms of polymyxin resistance have been identified in *A. baumannii*. The initial mechanism involves the modification of lipid A *via* phosphoethanolamine and/or galactosamine, orchestrated by the PmrAB two-component system. Mutations at single nucleotide levels or elevated expression of pmrA (response regulatory protein) or PmrB (histidine kinase sensor) trigger the upregulation of pmrC, subsequently activating the production of phosphoethanolamine transferase (PEtN). This enzyme alters lipid A structure [42–46]. Other genes influencing LPS biosynthesis and lipid A configuration have also been documented. Additionally, the involvement of efflux pumps in colistin resistance cannot be dismissed [47, 48].

Another *A. baumannii* polymyxin resistance mechanism involves the complete loss of LPS from the outer membrane, which stems from mutations or inactivation due to the insertion of the ISAba11 insertion sequence into the lpxA, lpxC, and lpxD genes. These genes encode enzymes accountable for the initial stages of polymyxin LPS biosynthesis [43, 44, 49].

Mutations within the gene responsible for glycosyltransferase, a component involved in LPS biosynthesis, have also been linked to polymyxin resistance [50, 51]. According to current literature, both resistance mechanisms negate polymyxin-triggered bacterial death by obstructing the interaction of polymyxins with OM. The mechanisms are governed by the pmrCAB operon (for lipid A modification with PEtN), naxD (for galactosamine modification), or the lpx biosynthetic cluster (for LPS loss) [42, 44–46].

The outer membrane lipoprotein VacJ is an integral part of the Vps-VacJ ABC transporter system, responsible for maintaining the presence of phospholipids and LPS within the outer membrane [52]. Mutations within the vacJ and pldA genes could contribute to *A. baumannii's* colistin resistance due to their role in preserving the asymmetrical lipid distribution in the outer membrane [53]. In 2016, the discovery of the plasmid-borne mcr-1 gene marked the first instance of a colistin-resistant gene with horizontal transmission capability [54]. Unlike its predecessors, this gene can be disseminated via plasmids, expanding the reach of colistin resistance [55]. In subsequent years, the mcr-4.3 gene variant, carried by a plasmid, has also been identified [56–58]. Understanding the intricacies of polymyxin resistance mechanisms has become imperative for maintaining the effectiveness of this antibiotic until novel therapeutic alternatives are available. Nevertheless, assessing susceptibility to polymyxins remains a contentious issue as numerous laboratories do not employ the microdilution technique recommended for this evaluation [59].

2.5 Heteroresistence

Heteroresistance refers to the emergence of resistance to a specific antibiotic within a population initially sensitive to that antibiotic based on *in vitro* susceptibility test cutoff points [60]. Some studies describe this phenomenon without specifying the antibiotic concentration range. In contrast, others identify heteroresistance when subpopulations of an isolate grow at concentrations exceeding minimum inhibitory concentration (MIC) values found in susceptibility tests yet still within the susceptibility range [61, 62]. This variability in definitions, detection methods, and prevalence complicates understanding of heteroresistance's clinical significance [63]. This phenotype might represent a natural progression of antibiotic resistance, allowing bacteria to grow in the presence of antibiotics following resistance acquisition by most of the microbial population [63]. In 2006, Li et al. [61] first reported heteroresistance to colistin in multidrug-resistant *A. baumannii* isolates, defining it as the emergence of resistance within a subpopulation of an otherwise susceptible (MIC ≤2 mg/L) group. Since then, this phenomenon has been widely observed, with prevalence ranging from 1.84–100% [64–66]. A related study showed higher heteroresistance in patients previously treated with colistin, suggesting prior colistin therapy might induce heteroresistance [64]. Additionally, the synergistic activity of colistin has been compromised when tested in antimicrobial combinations against heteroresistant carbapenem-resistant *A. baumannii* strains [67]. Findings regarding resistance stability within surviving subpopulations under nonselective conditions have varied across studies, implying a potential species-specific influence [60, 61, 64, 68]. Under colistin exposure, a subset of cells becomes colistin-dependent for optimal growth, indicating an adaptive response to colistin pressure and an intermediate stage between susceptibility or heteroresistance and full-blown colistin resistance [69, 70]. Hong et al. [60] found isolates displaying a heteroresistant phenotype at low antibiotic concentrations, distinct from the typical heteroresistant colistin isolates emerging at high colistin concentrations. The mechanisms of heteroresistance to colistin in *A. baumannii* are consistent with those previously described for colistin resistance, involving LpxACD, PmrCAB, and efflux pumps [60, 65, 68, 71, 72].

Detecting heteroresistant strains necessitates using the population profile analysis (PAP) method, the gold standard for identifying heteroresistance. In clinical practice, the introduction of the mini-PAP method, particularly for colistin with MIC >2 mg/L, has been recommended [73]. However, the fact that conventional susceptibility testing categorizes heteroresistant isolates as susceptible to colistin poses a notable concern [65]. Heteroresistance can sometimes be indicated by colonies within the growth inhibition zone, as seen with Etest® strips or disc diffusion assays. Nevertheless, standard dilution methods used for MIC determination fail to detect heteroresistance, potentially leading to suboptimal patient dosages. This suboptimal treatment might inadvertently select the resistant population, contributing to therapeutic failures [26, 74]. Inappropriate colistin use also holds significant potential for rapid resistance development and therapeutic inefficacy [75]. Under selection pressure, a subpopulation of resistant cells within a heteroresistant population can become predominant, yielding an entirely resistant population [68].

2.6 Clinical use

In clinical practice, polymyxins are employed as either polymyxin B or colistin. Despite their structural similarity, these drugs differ in their administered forms and exhibit distinct clinical pharmacokinetics (PK) [30]. Polymyxin B is directly administered in its active form as polymyxin B sulfate salt. In contrast, colistin is administered as an inactive prodrug called colistin metasulfate or colistimethate (CMS). Once metabolized, CMS is converted into the active ingredient colistin base. CMS is less toxic than colistin, and its conversion to colistin occurs gradually, coupled with rapid renal elimination.

Consequently, only about 20–25% of the administered CMS is effectively transformed into colistin [76–78]. Polymyxin B administration leads to quicker attainment of target concentrations [79]. Although polymyxin B and colistin exhibit comparable in vitro antimicrobial activity [30], differences in their plasma concentration profiles following therapy initiation will likely significantly impact their pharmacodynamic responses in patients.

3. Polymyxin toxicity

The 1990s saw the emergence of multidrug-resistant bacteria, including those resistant to β-lactams, aminoglycosides, and quinolones, causing nosocomial infections, particularly in intensive care units [80–83]. This scenario increased interest in polymyxins and spurred several reviews [84, 85]. These drugs' most significant adverse effects include nephrotoxicity, particularly acute renal failure, and neurotoxicity. The latter is thought to result from the high binding affinity of polymyxins to brain and renal tissues [86]. Additional effects encompass allergies leading to skin lesions resembling urticaria, pain at the injection site (with intramuscular administration), thrombophlebitis (with intravenous injection), fever, and eosinophilia [87, 88].

3.1 Nephrotoxicity

Nephrotoxicity ranks as the foremost adverse event often linked to the use of polymyxins. Thus, comprehending the mechanisms and risk factors for its development has been a focal point of research [89, 90]. Clinical manifestations of polymyxin-associated nephrotoxicity include direct toxicity to renal tubules leading to tubular necrosis, oxidative damage, decreased glomerular filtration rate, reduced creatinine clearance, and elevated serum urea and creatinine levels [80, 91]. Risk factors for kidney damage among polymyxin users encompass high doses, concurrent use of other nephrotoxic drugs, vasoactive medication requirements, and a higher body mass index [92–95]. The substantial concern with nephrotoxicity lies in its dose-dependent nature. In other words, the choice of therapy can influence the extent of drug-induced toxicity, potentially exacerbating the clinical condition of patients [96]. Dose-dependent nephrotoxicity is the most frequently reported adverse event with intravenous polymyxin use, affecting between 30 and 60% of patients [78, 85, 97–101]. However, it is often reversible [102]. While most studies have examined colistin, fewer studies have focused on polymyxin B. Due to the slower conversion of CMS to colistin, reaching therapeutic serum levels may be delayed, necessitating higher initial CMS doses to achieve effective treatment early on. However, this strategy is constrained by the potential for nephrotoxicity. Polymyxin B, administered directly in its active form, reaches the desired plasma concentration more promptly [30]. Recent literature suggests greater nephrotoxicity with colistin compared to polymyxin [103]. However, these findings require careful evaluation due to many factors influencing nephrotoxicity development, especially during the initial stages. Additionally, the potential nephrotoxicity of low polymyxin B doses may have been underestimated. Several studies have explored the efficacy of polymyxin B and colistin against *A. baumannii*, providing data on nephrotoxicity incidence and mortality (**Table 1**).

Acute kidney injury (AKI) is a prevalent clinical complication observed primarily in critical and hospitalized patients, characterized by the release of measurable proteins in both plasma and urine. This condition is rooted in the sudden decline of renal function, classified into risk, damage, failure, loss, and AKI stages [137, 138]. Critically ill patients suffering from AKI often face elevated mortality rates. This acute injury can progress to chronic kidney disease, defined by kidney damage and a glomerular filtration rate below 60 $mL/min/1.73m^2$ over 3 months. Therefore, discontinuing polymyxin therapy is imperative whenever signs of renal failure are detected. Supportive care, including monitoring fluid intake, output, and electrolytes, becomes necessary when renal dysfunction is associated with polymyxin use [85].

N° of patients/ therapy	GNB (n)	Definition of nephrotoxicity	Nephrotoxicity (%)	Mortality rate (%)	Ref.
60/COL	AB (39) PA (21)	CrL of 1.5 mg/dL or urea level of 50 mg/dL	27 (NRF) 58 (ABCL)	37	[104]
21/IVCOL	AB (21)	SCr value of 12 mg/dL, reduction in the calculated CLCr of 50% relative to the matter at antibiotic therapy initiation, or a decline in RF that resulted in the need for RRT	24	61.9	[105]
60/PB	AB (46) PA (2) AB + PA (2) NI (10)	Double the SCr for a value ≥2 mg/dL	14	20 57 (DRF) 15 (NDRF)	[106]
26/COL	PA (20) AB (6)	ND	14.4	33.3	[107]
16/IVCOL, AEROPB + AA	AB (16) PA (12)	Doubling of SCr	6	21 (EOT) 48 (AD)	[108]
19/IVCOL	PA (12) AB (5)	CrV at the beginning of COLtreatment was compared with the maximum value of creatinine during therapy as well as with the CrV at the end of treatment using a non-parametric test (Wilcoxon)	0	41.2	[109]
55/COL	AB (36) PA (19)	SCr value of 12 mg/dL, reduction in the calculated CLCr of 50% relative to the matter at antibiotic therapy initiation, or a decline in RF that resulted in the need for RRT	0	27	[110]
43/COL	PA (35) AB (8)	Acute RF was defined as a rise of 2 mg / dL in the SrCr level of patients with previously normal renal function	62.5	27.9	[111]
51/COL	AB (28) PA (23)	Normal renal function was defined as a SCr level of 1.3 mg/dl or lower.	8	24	[112]
37/IVPB, PBVN, both (IPB/PBVN), DOXI	AB (37)	Increase in SCr of 0.5 mg/dL, or increase ≥50% in SCr or reduction of ClCr ≥50%	21/6	27	[113]
45/IVPB	PA (20) AB (19) PA + AB (2) NI (4)	Acute increase in SCr level by >0.5 mg/dL over 24 h	4	52 (IH)	[114]
16/PB	PA (8) AB (5) KP (3) EC (1)	Increase in SCr of 0.5 mg/dL or a 50% reduction in CLCr	55	63	[98]

N° of patients/ therapy	GNB (n)	Definition of nephrotoxicity	Nephrotoxicity (%)	Mortality rate (%)	Ref.
82/COL, PB	AB (82)	Doubling of SCr (any time during treatment compared with the start of therapy) or increase by 1 mg/dL if initial SCr was 1.4 mg/dL	26 (COl group) 27 (PB group)	56 (COL group) 61 (PB group)	[115]
114/IVPB	PA (95) AB (13) KP (1) PA + AB (2) NI (3)	Baseline SCr < 1.5 mg/dL when SCr levels increased to 1.8 mg/dL (AKI) or baseline SCr 1.5 mg/dL when SCr levels increased to >50%, or there was a need for dialysis	22 AKI/NS	61.4 92 (DAKI) 53 (NDAKI)	[116]
276/PB	PA (126) AB (86) NI (64)	MRI: 50% but <100% (increase in creatinine concentration during therapy); MORI: 100% (increase in creatinine concentration but with no need for hemodialysis); SRI: need for hemodialysis during therapy	15.7 (MRI) 38.3 (MOSRI)	60.5 (IH)	[99]
80/PB (NPD or CD)	KP (49) AB (21) PA (14) EC (4) ECO (1)	Defined by RIFLE criteria	40 (1 week after the last dose)	15 vs. 20 (EOT) 30 vs. 38 (EOH)	[116]
173/COL, PB	AB (107) PA (46)	Defined by RIFLE criteria	60 (COL group) 41.8 (PB group)	ND	[92]
32/IVPB	AB (26) PA (1) ECO (1) SE (1) Mu (3)	Defined by RIFLE criteria	18.7	28.1 (EOT)	[117]
225/IVCOL, PB	PA (103) AB (74) KP (52) ECO (11) Other (17)	Prevalence of nephrotoxicity within 30 days in colistimethate group compared with PB group Comparison of nephrotoxicity prevalence in matched patients	21.4 (COL group) 21.4 (PB group)	55.3 (COL group) 21.1 (PB group)	[93]
104/PB	AB (34) KP (25) PA (11) Mu (34)	Defined by RIFLE criteria	14.4	47	[118]
132/COL, PB	AB (43) PA (22) KP (12) DI (18) NI (37)	Classified according to AKIN criteria	20.8 (AKI/PB group) 38.9 (AKI/COL group)	47	[119]
36/PB	A spp. (12) KP (8) PA (6) ECO (6) E spp. (5) Other (9)	Increase of 100% of SCr level from baseline	21.4	44.5	[120]

N° of patients/ therapy	GNB (n)	Definition of nephrotoxicity	Nephrotoxicity (%)	Mortality rate (%)	Ref.
410/PB	AB (150) PA (45) KP (42) ECO (5) EA (5) NI or NR (162)	Defined by RIFLE criteria	12.7	42	[121]
151/ PB	KP (92) AB (32) PA (17) Other (10)	AKI: increase in SCr 1.5 times the value at PB initiation or the initiation of RRT by day 7 of PB treatment, defined by RIFLE criteria	35.8 AKI	NS	[122]
192/ IVPB	KP (92) AB (53)	Defined by RIFLE criteria	45.8	NS	[123]
491/IVCOL, PB	AB (180) KP (55) PA (51) EA (9) ECO (5) NI (190)	Incidence of AKI by RIFLE criteria	38.3 (COL group) 12.7 (PB group)	NS	[124]
291/PB, NVPT, in vitro VCT	AB (228) PA (61) KP (14) Other (7)	Defined by RIFLE criteria	98 of 291	23	[125]
112/IVCOL, PB	KP (31) AB (22) PA (19) ECO (5) NI (35)	A two-fold increase in SCr or a 50% decrease in estimated CLCr	26.8	NS	[103]
84/IVPB, PBM, PB/ CARB, CEFO/ SUL	AB (81)	MRI: decrease in baseline CLCr of 50% or doubling of baseline SCr in patients with normal renal function, or an increase of baseline SCr of 50% or decrease of CLCr of 20% in patients with abnormal baseline anal function	7.1 (RI)	48.8 (IH)	[126]
222/PB	AB (67) E (50) PA (15) Other (4) NI (86)	Defined by RIFLE criteria	46.3	60.3	[127]
273/PB	KP (108) PA (74) AB (77) ECO (22) Other (9)	Defined by RIFLE criteria	32	47 (ODD) 17 (TDD)	[128]

N° of patients/ therapy	GNB (n)	Definition of nephrotoxicity	Nephrotoxicity (%)	Mortality rate (%)	Ref.
183/IVCOL or ICOL, IVCOL/ ICOL	*Acinetobacter calcoaceticus-Acinetobacter baumannii* (Acb) complex (183)	Increase in SCr of ≥0.3 mg/ dL in 2 days or ≥ 50% in 7 days after COL treatment without other defined causes	13.3	19.1	[129]
250/COL + MERO	AB (197) AB+KP (1) NS (52)	Classified according to AKI criteria	30.8	41.6	[130]
39/IVCOL	PA (34) AB (5) EC (1)	Based on the ROC curve, the cutoff value of the colistin trough concentration that would predict nephrotoxicity was 2.02 mg/mL	47.6	33.3	[131]
87/COL	AB (73) NS (14)	Increase in the SCr level by at least 50% from the baseline after≥48 h	27.6	NS	[132]
50/COL		Defined by RIFLE criteria	54 (MIC ≤0.5 µg/mL	NS	[133]
25/IVCOL	AB (25)	Increase in SCr to ≥1.5- fold from baseline, decrease in the estimated CLCr to <75% from baseline, or requirement for RRT	20	40 (IH)	[134]
163/COL	A spp. (118) PA (32) KP (7) E spp. (6)	Followed by KDIGO classification: creatinine elevation of ≥0.3 mg/dL in 48 h or ≥ 1.5 times baseline creatinine in an interval of up to 7 days	46	17.8	[135]
101/COL	AB (101)	Defined by RIFLE criteria	52.6 (LD group) 20.5 (WLD group)	51.3	[136]

COL, colistin; PB, polymyxin; PBM, polymyxin B monotherapy; IVCOL, intravenously colistin; ICOL, inhaled colistin; AEROPB, aerosolized polymyxin B; IVPB, intravenously polymyxin B; TDD, twice daily dosing; NRF, normal renal function; ABCL, abnormal baseline creatinine levels; PBVN, polymyxin B via nebulization; NPD, new protocol design; CD, conventional dosing; NVPT, nonvalidated polymyxin therapy; VPCT, validated polymyxin combination therapy; CARB, carbapenems; CEFO, cefoperazone; SUL, sulbactam; DOXI, doxycycline; MERO, meropenem; AB, Acinetobacter baumannii; PA, Pseudomonas aeruginosa; KP, Klebsiella pneumoniae; EC, Enterobacter cloacae; ECO, Escherichia coli; E spp., Enterobacter spp.; EA, Enterobacter aerogenes; A spp., Acinetobacter spp.; E, Enterobacteriaceae; DI, dual infection; SCr, serum creatinine; CLCr, creatinine clearance; CrL, Creatinine level; CrV, Creatinine values; NI, none identified; MRI, mild renal impairment; NR, not request; NS, not stated; NRF, normal renal function; ABCL, abnormal baseline creatinine levels; DRF, Developed renal failure; NDRF, not developed renal falure; Mu, multiple; AERO, aerosolized; RI, Renal impairment; RF, renal failure; AD, at discharge; AA, antimicrobial agente; ND, not determined; EOT, end of treatment; IH, In-hospital; RIFLE, Risk, Injury, Failure, Loss of kidney function and End-stage kidney disease; EOH, End of hospitalization; AKI, acute kidney injury; AKIN, acute kidney injury network; ODD, Once daily dosing; TDD, twice daily dosing; DI, dual infection; MORI, moderate and severe renal impairment; DAKI, developed Aki; NDAKI, not developed AKI; RRT, renal replacement therapy; KDIGO, kidney disease improving global outcomes; LD, loading dose; WLD, without loading amount.

Table 1.
Studies report nephrotoxicity during polymyxin therapy against Acinetobacter baumannii.

3.2 Neurotoxicity

Neurotoxicity constitutes another undesirable consequence of polymyxin administration. Neurotoxicity related to polymyxins affects 7–27% of patients, with most cases involving concurrent renal failure [139, 140]. Symptoms of neurotoxicity encompass weakness, peripheral and facial paresthesia, ataxia, ophthalmoplegia, nystagmus, difficulty swallowing, and eyelid ptosis [88, 139–144]. Severe manifestations include muscle blockade leading to respiratory failure, often requiring ventilatory support for 10 to 48 hours [140, 141]. Typically, these symptoms decrease upon tapering or discontinuation of the drug. The administration of colistin triggers the activation of pro-inflammatory mediators within neuronal cells [145]. Research indicates that neurotoxicity entails a complex interplay of apoptotic and inflammatory pathways. Studies involving colistin treatment (15 mg/kg/day for 7 days) revealed significant mitochondrial dysfunction in central and peripheral nervous tissues [146, 147]. Similarly, exposure to colistin (200 μM/24 h) induced apoptosis in around 50% of neuronal N2a cells in mice [145]. Further exploration using Western blotting and immunohistochemistry demonstrated that colistin-induced apoptosis in N2a neuronal cells hinges on generating reactive oxygen species (ROS) and the mitochondrial pathway [145, 148, 149]. Interestingly, co-administration of neuroprotective agents, such as curcumin and minocycline demonstrated, *in vivo* efficacy against polymyxin-induced neurotoxicity [145, 149].

3.3 Skin hyperpigmentation

Although nephrotoxicity ranks as polymyxin B's most significant adverse reaction, another substantial side effect is skin hyperpigmentation. Polymyxin B induces this condition, which impacts psychological well-being and results in significant esthetic harm [150–158]. Cutaneous hyperpigmentation has been observed as a reaction to polymyxin B, affecting adults and pediatric and neonatal patients [151, 153–155]. According to cohort studies, the incidence of cutaneous hyperpigmentation attributed to this drug ranges from 8–15% [151, 152]. Cutaneous hyperpigmentation involves biochemical and immunological mechanisms, primarily associated with histaminergic receptors that stimulate melanogenesis, ultimately leading to melanin deposition in the dermis [150]. Typically, skin darkening manifests between the third and seventh days following the commencement of intravenous polymyxin B treatment. This phenomenon does not show significant disparities concerning light exposure or infection sites across patients [152]. Hyperpigmentation is often concentrated on the face and neck regions with higher melanocyte density, while the rest of the body remains unaffected during treatment [152, 154, 155, 159].

In some cases, discontinuing polymyxin B treatment reveals hyperpigmentation that can persist for months [150]. During the COVID-19 pandemic, polymyxin B treatment was administered to physicians with COVID-19 and secondary multidrug-resistant bacterial infections, resulting in hyperpigmentation on the head and neck [160]. This pigmentary disorder may be associated with AKI in critically ill COVID-19 patients [160]. Excessive accumulation of polymyxin B might contribute to aberrant hyperpigmentation in neonates and infants with immature renal function [153, 158].

4. Conclusions

In summary, this chapter presents a comprehensive review of the toxicity of polymyxins, which serve as the last resort for treating infections caused by carbapenem-resistant *A. baumannii*. The chapter begins by highlighting the current significance of *A. baumannii* as a challenging pathogen in healthcare settings, given its formidable ability to develop resistance through diverse mechanisms. Accordingly, it ranks as a high-priority microorganism for research and developing new antimicrobials. Despite their notable toxicity, polymyxins were re-introduced in the late 1990s due to escalating carbapenem resistance and limited alternative options. The chapter delves into the discovery and isolation of polymyxins, focusing on polymyxin B and polymyxin E (colistin) as the two varieties in clinical use. Their distinctive structural features enable interactions with cell membrane LPS, leading to membrane disruption through the cationic peptide ring's hydrophilic nature and the fatty acyl chain's hydrophobic characteristics. The emergence of polymyxin resistance is addressed, focusing on its occurrence through mutation or adaptation in Gram-negative bacteria. In *A. baumannii*, the resistance mechanism involves genes influencing LPS biosynthesis and lipid A structure.

Additionally, efflux pumps and the mcr-1 gene contribute to colistin resistance. The phenomenon of heteroresistance to colistin in *A. baumannii* is explored, emphasizing its reliance on the population profile analysis method for detection. This method, recognized as the gold standard, has revealed the presence of heteroresistance and its association with the previously discussed resistance mechanisms. Lastly, the clinical use of polymyxin B and colistin is outlined alongside their toxic effects. Nephrotoxicity is a prominent adverse event tied to polymyxin use, characterized by direct renal tubule toxicity and dose-dependent, often reversible effects. Most studies focus on colistin. One of its clinical complications is acute kidney injury (AKI). Neurotoxicity emerges as another unwanted effect, causing symptoms that generally wane with drug reduction or discontinuation. Severe cases might involve muscle blockade leading to respiratory failure. Furthermore, skin hyperpigmentation, a recognized reaction to polymyxin B, affects patients of varying ages through complex biochemical and immunological mechanisms.

Acknowledgements

This research was funded by the Carlos Chagas Filho Foundation for Research Support of the State of Rio de Janeiro/FAPERJ (#110.198-13) and the Brazilian Council for Scientific Research (CNPq, #467.488/2014-2 and 301744/2019-0). Funding was also provided by FAPERJ (#210.003/2018) through the National Institutes of Science and Technology Program (INCT) to Carlos M. Morel (INCT-IDPN).

Author contributions

Conceived and designed the experiments: K.R., T.P.G.C.; writing—original draft: K.R.; review and editing: K.R., S.G.D.-S.; funding: S.G.D.-S. All authors have read and agreed to the published version of the manuscript.

Conflict of interest

The authors declare no conflict of interest.

Author details

Karyne Rangel[1,2*], Thiago Pavoni Gomes Chagas[3,4] and
Salvatore Giovanni De-Simone[1,2,5*]

1 Center for Technological Development in Health (CDTS), National Institute of Science and Technology for Innovation in Neglected Population Diseases (INCT-IDPN), Oswaldo Cruz Foundation (FIOCRUZ), Rio de Janeiro, Brazil

2 Laboratory of Epidemiology and Molecular Systematics (LESM), Oswaldo Cruz Institute, Oswaldo Cruz Foundation (FIOCRUZ), Rio de Janeiro, Brazil

3 Molecular Epidemiology and Biotechnology Laboratory, Department of Pathology, Medical School, Federal Fluminense University, Niterói, Brazil

4 Faculty of Pharmacy, Rodolpho Albino Laboratory, Federal Fluminense University, Niterói, Brazil

5 Post-Graduation Program in Science and Biotechnology, Department of Molecular and Cellular Biology, Biology Institute, Federal Fluminense University (UFF), Niterói, Brazil

*Address all correspondence to: karyne.rangelk@gmail.com and salvatore.simone@fiocruz.br

References

[1] Tang KWK, Millar BC, Moore JE. Antimicrobial Resistance (AMR). British Journal of Biomedical Science. 2023;**80**:11387. DOI: 10.3389/bjbs.2023.11387

[2] Antimicrobial Resistance Collaborators. Global burden of bacterial antimicrobial resistance in 2019: A systematic analysis. Lancet. 2022;**399**(10325):629-655. DOI: 10.1016/S0140-6736(21)02724-0

[3] Thompson T. The staggering death toll of drug-resistant bacteria. Nature. 2022. DOI: 10.1038/d41586-022-00228-x

[4] Wong D, Nielsen TB, Bonomo RA, Pantapalangkoor P, Luna B, Spellberg B. Clinical and pathophysiological overview of *Acinetobacter* infections: A century of challen ges. Clinical Microbiology Reviews. 2017;**30**(1):409-447. DOI: 10.1128/CMR.00058-16

[5] Ibrahim S, Al-Saryi N, Al-Kadmy IMS, Aziz SN. Multidrug-resistant *Acinetobacter baumannii* is an emerging concern in hospitals. Molecular Biology Reports. 2021;**48**(10):6987-6998. DOI: 10.1007/s11033-021-06690-6

[6] Hamidian M, Nigro SJ. Emergence, molecular mechanisms and global spread of carbapenem-resistant *Acinetobacter baumannii*. Microbial Genomics. 2019;**5**(10):e000306. DOI: 10.1099/mgen.0.000306

[7] Ramirez MS, Bonomo RA, Tolmasky ME. Carbapenemases: Transfor ming *Acinetobacter baumannii* into a yet more dangerous menace. Biomolecules. 2020;**10**(5):720. DOI: 10.3390/biom10050720

[8] Tacconelli E, Carrara E, Savoldi A, Harbarth S, Mendelson M, Monnet DL, et al. WHO pathogens priority list working group. Discovery, research, and development of new antibiotics: The WHO priority list of antibiotic-resistant bacteria and tuberculosis. The Lancet Infectious Diseases. 2018;**18**(3):318-327. DOI: 10.1016/S1473-3099(17)30753-3

[9] Ellis RC, Roberts EK, Grier JT, Fiester SE. *Acinetobacter baumannii* infections that are resistant to treatment: Warning signs from the COVID-19 pandemic. Future Microbiology. 2022;**17**:1345-1347. DOI: 10.2217/fmb-2022-0153

[10] Magiorakos AP, Srinivasan A, Carey RB, Carmeli Y, Falagas ME, Giske CG, et al. Multidrug-resistant, extensively drug-resistant and pan drug-resistant bacteria: An international expert proposal for interim standard definitions for acquired resistance. Clinical Microbiology and Infection. 2012;**18**(3):268-281. DOI: 10.1111/j.1469-0691.2011.03570.x

[11] Piperaki ET, Tzouvelekis LS, Miriagou V, Daikos GL. Carbapenem-resistant *Acinetobacter baumannii*: In pursuit of an effective treatment. Clinical Microbiology and Infection. 2019;**25**(8):951-957. DOI: 10.1016/j.cmi.2019.03.014

[12] Weinberg SE, Villedieu A, BagdasarianN,KarahN,TeareL,ElaminWF. Control and management of multidrug-resistant *Acinetobacter baumannii*: A review of the evidence and proposal of novel approaches. Infection Prevention in Practice. 2020;**2**(3):100077. DOI: 10.1016/j.infpip.2020.100077

[13] Nang SC, Azad MAK, Velkov T, Zhou QT, Li J. Rescuing the last-line polymyxins: Achievements and

challenges. Pharmacological Reviews. 2021;**73**(2):679-728. DOI: 10.1124/pharmrev.120.000020

[14] Poirel L, Jayol A, Nordmann P. Polymyxins: Antibacterial activity, susceptibility testing, and resistance mechanisms encoded by plasmids or chromosomes. Clinical Microbiology Reviews. 2017;**30**(2):557-596. DOI: 10.1128/CMR.00064-16

[15] Mmatli M, Mbelle NM, Maningi NE, Osei Sekyere J. Emerging transcriptional and genomic mechanisms mediating carbapenem and polymyxin resistance in *Entero bacteriaceae*: A systematic review of current reports. mSystems. 2020;**5**(6):e00783-e00720. DOI: 10.1128/mSystems.00783-20

[16] Aslan AT, Akova M, Paterson DL. Next-generation polymyxin class of antibiotics: A ray of hope illuminating a dark road. Antibiotics (Basel). 2022;**11**(12):1711. DOI: 10.3390/antibiotics11121711

[17] Bayraktar I, Halacli B, Demirkan K, Topeli A. Polymyxin B-related neurotoxicity: A brief case report. European Journal of Hospital Pharmacy. 2023;7. DOI: 10.1136/ejhpharm-2023-003786

[18] Ainsworth GC, Brown AM, Brownlee G. Aerosporin, an antibiotic produced by *Bacillus aerosporus* Greer. Nature. 1947;**159**(4060):263. DOI: 10.1038/16 0263a0

[19] Benedict RG, Langlykke AF. Antibiotic activity of *Bacillus polymyxa*. Journal of Bacteriology. 1947;**54**(1):24

[20] Stansly PG, Shepherd RG, White HJ. Polymyxin: A new chemotherapeutic agent. Bulletin of the Johns Hopkins Hospital. 1947;**81**(1):43-54

[21] Brownlee G, Bushby SR. Chemotherapy and pharmacology of aerosporin; a selective gram-negative antibiotic. Lancet. 1948;**1**(6491):127-132. DOI: 10.1016/s0140-6736(48)90090-7

[22] Shepherd RG, Stansly PG, Winterbottom R, English JP, Fellows CE, Ananenko NH, et al. Chemical studies on polymyxin; isolation and preliminary purification. Journal of the American Chemical Society. 1948;**70**(11):3771-3774. DOI: 10.1021/ja01191a068

[23] Brownlee G. Antibiotics derived from bacillus polymyxin. Annals of the New York Academy of Sciences. 1949;**51**(Art. 5):875-878. DOI: 10.1111/j.1749-6632.1949.tb27313.x

[24] Jones TS. Chemical evidence for the multiplicity of the antibiotics produced by *Bacillus polymyxa*. Annals of the New York Academy of Sciences. 1949;**51**(Art. 5):909-916. DOI: 10.1111/j.1749-6632.1949.tb27317.x

[25] White HJ, Alverson CM, Baker MJ, Jackson ER. Comparative biological studies of polymyxin and aerosporin. Annals of the New York Academy of Sciences. 1949;**51**(Art. 5):879-890. DOI: 10.1111/j.1749-6632.1949.tb27314.x

[26] Landman D, Georgescu C, Martin DA, Quale J. Polymyxins revisited. Clinical Microbiology Reviews. 2008;**21**(3):449-465. DOI: 10.1128/CMR.00006-08

[27] Yu Z, Qin W, Lin J, Fang S, Qiu J. Antibacterial mechanisms of polymyxin and bacterial resistance. BioMed Research International. 2015;**2015**:679109. DOI: 10.1155/2015/679 109

[28] Stansly PG, Brownlee G. Nomenclature of polymyxin antibiotics. Nature. 1949;**163**(4146):611. DOI: 10.1038/163611a0

[29] Trimble MJ, Mlynárčik P, Kolář M, Hancock RE. Polymyxin: Alternative mechanisms of action and resistance. Cold Spring Harbor Perspectives in Medicine. 2016;**6**(10):a025288. DOI: 10.1101/cshperspect.a025288

[30] Roberts KD, Azad MA, Wang J, Horne AS, Thompson PE, Nation RL. Et. Antimicrobial activity and toxicity of the major lipopeptide components of polymyxin B and colistin: Last-line antibiotics against multidrug-resistant gram-negative bacteria. ACS Infectious Diseases. 2015;**1**(11):568-575. DOI: 10.1021/acsI nfecdis.5b00085

[31] Falagas ME, Rafailidis PI, Matthaiou DK. Resistance to polymyxins: Mechanisms, frequency and treatment options. Drug Resistance Updates. 2010;**13**(4-5):132-138. DOI: 10.1016/j.drup.2010.05.002

[32] Zavascki AP, Bulitta JB, Landersdorfer CB. Combination therapy for carbapenem-resistant gram-negative bacteria. Expert Review of Anti-Infective Therapy. 2013;**11**(12):1333-1353. DOI: 10.1586/14787210.2013.845523

[33] Gales AC, Jones RN, Sader HS. Contemporary activity of colistin and polymyxin B against a worldwide collection of gram-negative pathogens: Results from the SENTRY antimicrobial surveillance program (2006-09). The Journal of Antimicrobial Chemotherapy. 2011;**66**(9):2070-2074. DOI: 10.1093/jac/dkr239

[34] Groisman EA, Kayser J, Soncini FC. Regulation of polymyxin resistance and adaptation to low-Mg2+ environments. Journal of Bacteriology. 1997;**179**(22):7040-7045. DOI: 10.1128/jb.179.22.7040-7045.1997

[35] Zavascki AP, Goldani LZ, Li J, Nation RL. Polymyxin B for the treatment of multidrug-resistant pathogens: A critical review. The Journal of Antimicrobial Chemotherapy. 2007;**60**(6):1206-1215. DOI: 10.1093/jac/dkm357

[36] Mohapatra SS, Dwibedy SK, Padhy I. Polymyxins, the last-resort antibiotics: Mode of action, resistance emergence, and potential solutions. Journal of Biosciences. 2021;**46**(3):85. DOI: 10.1007/s12038-021-00209-8

[37] Skiada A, Markogiannakis A, Plachouras D, Daikos GL. Adaptive resistance to cationic compounds in *Pseudomonas aeruginosa*. International Journal of Antimicrobial Agents. 2011;**37**(3):187-193. DOI: 10.1016/j.ijantimicag.2010.11.019

[38] Trent MS, Ribeiro AA, Lin S, Cotter RJ, Raetz CR. An inner membrane enzyme in salmonella and *Escherichia coli* that transfers 4-amino-4-deoxy-L-arabinose to lipid A: Induction on polymyxin-resistant mutants and role of a novel lipid-linked donor. The Journal of Biological Chemistry. 2001;**276**(46):43122-43131. DOI: 10.1074/jbc.M1069612 00

[39] Olaitan AO, Morand S, Rolain JM. Mechanisms of polymyxin resistance: Acquired and intrinsic resistance in bacteria. Frontiers in Microbiology. 2014;**5**:643. DOI: 10.3389/fmicb.2014.00643

[40] Jeannot K, Bolard A, Plésiat P. Resistance to polymyxins in gram-negative organisms. International Journal of Antimicrobial Agents. 2017;**49**(5):526-535. DOI: 10.1016/j.ijantimicag.2016.11.029

[41] Arroyo LA, Herrera CM, Fernandez L, Hankins JV, Trent MS, Hancock RE. The pmrCAB operon mediates polymyxin resistance

phosphoethanolamine modification of lipid A. Antimicrobial Agents and Chemotherapy. 2011;**55**(8):3743-3751. DOI: 10.1128/AAC.00256-11 in *Acinetobacter baumannii* ATCC 17978 and clinical isolates through

[42] Moffatt JH, Harper M, Adler B, Nation RL, Li J, Boyce JD. Insertion sequence ISAba11 is involved in colistin resistance and loss of lipopolysaccharide in *Acinetobacter baumannii*. Antimicrobial Agents and Chemotherapy. 2011;**55**(6):3022-3024. DOI: 10.1128/AAC.01732-10

[43] Moffatt JH, Harper M, Harrison P, Hale JD, Vinogradov E, Seemann T, et al. Colistin resistance in *Acinetobacter baumannii* is mediated by complete loss of lipopolysaccharide production. Antimicrobial Agents and Chemotherapy. 2010;**54**(12):4971-4977. DOI: 10.11.28/AAC.00834-10

[44] Chin CY, Gregg KA, Napier BA, Ernst RK, Weiss DS. A PmrB-regulated deacetylase required for lipid a modification and polymyxin resistance in *Acinetobacter baumannii*. Antimicrobial Agents and Chemotherapy. 2015;**59**(12):7911-7914. DOI: 10.1128/AAC.00515-15

[45] Park YK, Choi JY, Shin D, Ko KS. Correlation between overexpression and amino acid substitution of the PmrAB locus and colistin resistance in *Acinetobacter baumannii*. International Journal of Antimicrobial Agents. 2011;**37**(6):525-530. DOI: 10.1016/j.ijantimicag.2011.02.008

[46] Da Silva GJ, Domingues S. Interplay between colistin resistance, virulence and fitness in *Acinetobacter baumannii*. Antibiotics (Basel). 2017;**6**(4):28. DOI: 10.3390/antibiotics6040028

[47] Lin MF, Lin YY, Lan CY. Contribution of EmrAB efflux pumps to colistin resistance in *Acinetobacter baumannii*. Journal of Microbiology. 2017;**55**(2):130-136. DOI: 10.1007/s12275-017-6408-5

[48] Girardello R, Visconde M, Cayô R, Figueiredo RC, Mori MA, Lincopan N, et al. Diversity of polymyxin resistance mechanisms among *Acinetobacter baumannii* clinical isolates. Diagnostic Microbiology and Infectious Disease. 2017;**87**(1):37-44. DOI: 10.1016/j.diagmicrobio.2016.10.011

[49] Hood MI, Becker KW, Roux CM, Dunman PM, Skaar EP. Genetic determinants of intrinsic colistin tolerance in *Acinetobacter baumannii*. Infection and Immunity. 2013;**81**(2):542-551. DOI: 10.1128/IAI.00704-12. Erratum in: Infect Immun. 2014 Jan; 82 (1):469

[50] Lean SS, Yeo CC, Suhaili Z, Thong KL. Comparative genomics of two ST 195 carbapenem-resistant *Acinetobacter baumannii* with different susceptibility to polymyxin revealed underlying resistance mechanism. Frontiers in Microbiology. 2016;**6**:1445. DOI: 10.3389/fmicb.2015.01445

[51] Malinverni JC, Silhavy TJ. An ABC transport system that maintains lipid asymmetry in the gram-negative outer membrane. Proceedings of the National Academy of Sciences of the United States of America. 2009;**106**(19):8009-8014. DOI: 10.1073/pnas.0903229106

[52] Thi Khanh Nhu N, Riordan DW, Do Hoang Nhu T, Thanh DP, Thwaites G, Huong Lan NP, et al. The induction and identification of novel colistin resistance mutations in *Acinetobacter baumannii* and their implications. Scientific Reports. 2016;**22**(6):28291. DOI: 10.1038/srep28291

[53] Liu YY, Wang Y, Walsh TR, Yi LX, Zhang R, Spencer J, et al. Emergence

of plasmid-mediated colistin resistance mechanism MCR-1 in animals and human beings in China: A microbiological and molecular biological study. The Lancet Infectious Diseases. 2016;**16**(2):161-168. DOI: 10.1016/S1473-3099(15)00 424-7

[54] Paterson DL, Harris PN. Colistin resistance: A major breach in our last line of defense. The Lancet Infectious Diseases. 2016;**16**(2):132-133. DOI: 10.1016/S1473-3099(15) 00463-6

[55] Ma F, Shen C, Zheng X, Liu Y, Chen H, Zhong L, et al. Identification of a novel plasmid carrying *mcr-4.3* in an *Acinetobacter baumannii* strain in China. Antimicrobial Agents and Chemotherapy. 2019;**63**(6):e00133-e00119. DOI: 10.1128/AAC.00133-19

[56] Bitar I, Medvecky M, Gelbicova T, Jakubu V, Hrabak J, Zemlickova H, et al. Complete nucleotide sequences of *mcr-4.3*-carrying plasmids in *Acinetobacter baumannii* sequence type 345 of human and food origin from the Czech Republic, the first case in Europe. Antimicrobial Agents and Chemotherapy. 2019;**63**(10):e01166-e01119. DOI: 10.1128/AAC.01166-19

[57] Martins-Sorenson N, Snesrud E, Xavier DE, Cacci LC, Iavarone AT, McGann P, et al. A novel plasmid-encoded mcr-4.3 gene in a colistin-resistant *Acinetobacter baumannii* clinical strain. The Journal of Antimicrobial Chemotherapy. 2020;**75**(1):60-64. DOI: 10.1093/jac/dkz413

[58] Hindler JA, Humphries RM. Colistin MIC variability by method for contemporary clinical isolates of multidrug-resistant gram-negative bacilli. Journal of Clinical Microbiology. 2013;**51**(6):1678-1684. DOI: 10.1128/JCM.03385-12

[59] Hong YK, Kim H, Ko KS. Two types of colistin heteroresistance in *Acinetobacter baumannii* isolates. Emerging Microbes & Infections. 2020;**9**(1):2114-2123. DOI: 10.1080/22221751.2020.1821584

[60] Li J, Rayner CR, Nation RL, Owen RJ, Spelman D, Tan KE, et al. Heteroresistance to colistin in multidrug-resistant *Acinetobacter baumannii*. Antimicrobial Agents and Chemotherapy. 2006;**50**(9):2946-2950. DOI: 10.1128/AAC.00103-06

[61] Lo-Ten-Foe JR, de Smet AM, Diederen BM, Kluytmans JA, van Keulen PH. Comparative evaluation of the VITEK 2, disk diffusion, e-test, broth microdilution, and agar dilution susceptibility testing methods for colistin in clinical isolates, including heteroresistant *Enterobacter cloacae* and *Acinetobacter baumannii* strains. Antimicrobial Agents and Chemotherapy. 2007;**51**(10):3726-3730. DOI: 10.1128/AAC.01406-06

[62] Falagas ME, Makris GC, Dimopoulos G, Matthaiou DK. Heteroresistance: A concern of increasing clinical significance? Clinical Microbiology and Infection. 2008;**14**(2):101-104. DOI: 10.1111/j.1469-0691.2007.01912.x

[63] Hawley JS, Murray CK, Jorgensen JH. Colistin heteroresistance in *Acinetobacter* and its association with previous colistin therapy. Antimicrobial Agents and Chemotherapy. 2008;**52**(1):351-352. DOI: 10.1128/AAC.00766-07

[64] Chen L, Lin J, Lu H, Zhang X, Wang C, Liu H, et al. Deciphering colistin heteroresistance in *Acinetobacter baumannii* clinical isolates from Wenzhou, China. Journal of Antibiotics (Tokyo). 2020;**73**(7):463-470. DOI: 10.1038/s4142 9-020-0289-2

[65] Karakonstantis S, Saridakis I. Colistin heteroresistance in *Acinetobacter spp.*: Systematic review and meta-analysis of the prevalence and discussion of the mechanisms and potential therapeutic implications. International Journal of Antimicrobial Agents. 2020;**56**(2):106065. DOI: 10.1016/j.ijantimicag.2020.106065

[66] Rodriguez CH, De Ambrosio A, Bajuk M, Spinozzi M, Nastro M, Bombicino K, et al. In vitro antimicrobials activity against endemic *Acinetobacter baumannii* multiresistant clones. Journal of Infection in Developing Countries. 2010;**4**(3):164-167. DOI: 10.3855/jidc.604

[67] Machado D, Antunes J, Simões A, Perdigão J, Couto I, McCusker M, et al. Contribution of efflux to colistin heteroresis tance in a multidrug-resistant *Acinetobacter baumannii* clinical isolate. Journal of Medical Microbiology. 2018;**67**(6):740-749. DOI: 10.1099/jmm.0.000741

[68] Kon H, Hameir A, Temkin E, Keren-Paz A, Schwartz D, Schechner V, et al. Colistin dependency among colistin-heteroresistant *Acinetobacter baumannii* isolates. Microorganisms. 2021;**10**(1):58. DOI: 10.3390/microorganis ms 10010058

[69] Lee JY, Chung ES, Ko KS. Transition of colistin dependence into colistin resistance in *Acinetobacter baumannii*. Scientific Reports. 2017;**7**(1):14216. DOI: 10.1038/s415 98-017-14609-0

[70] Jovcic B, Novovic K, Dekic S, Hrenovic J. Colistin resistance in environmental isolates of *Acinetobacter baumannii*. Microbial Drug Resistance. 2021;**27**(3):328-336. DOI: 10.1089/mdr.2020.0188

[71] Charretier Y, Diene SM, Baud D, Chatellier S, Santiago-Allexant E, van Belkum A, et al. Colistin heteroresistance and involvement of the PmrAB regulatory system in *Acinetobacter baumannii*. Antimicrobial Agents and Chemotherapy. 2018;**62**(9):e00788-e00718. DOI: 10.1128/AAC.00788-18

[72] Yau W, Owen RJ, Poudyal A, Bell JM, Turnidge JD, Yu HH, et al. Colistin hetero-resistance in multidrug-resistant *Acinetobacter baumannii* clinical isolates from the Western Pacific region in the SENTRY antimicrobial surveillance program. The Journal of Infection. 2009;**58**(2):138-144. DOI: 10.1016/j.jinf.2008.11.002

[73] Martínez-Martínez L. Muerte bacteriana y heterorresistencia a los antimicrobia nos [bacterial death and heteroresistance to antimicrobial agents]. Enfermedades Infecciosas y Microbiología Clínica. 2008;**26**(8):481-484. DOI: 10.1016/s0213-005x(08)72 774-5

[74] Li J, Rayner CR, Nation RL, Deans R, Boots R, Widdecombe N, et al. Pharmacokinetics of colistin methanesulfonate and colistin in a critically ill patient receiving continuous venovenous hemodiafiltration. Antimicrobial Agents and Chemotherapy. 2005;**49**(11):4814-4815. DOI: 10.1128/AAC.49.11.4814-4815.2005

[75] PlachourasD, KarvanenM, FribergLE, Papadomichelakis E, Antoniadou A, TsangarisI, et al. Populationpharmacokinetic analysis of colistin methanesulfonate and colistin after intravenous administration in critically ill patients with infections caused by gram-negative bacteria. Antimicrobial Agents and Chemotherapy. 2009;**53**(8):3430-3436. DOI: 10.1128/AAC.01361-08

[76] Garonzik SM, Li J, Thamlikitkul V, Paterson DL, Shoham S, Jacob J, et al.

Population pharmacokinetics of colistin methanesulfonate and formed colistin in critically ill patients from a multicenter study provide dosing suggestions for various categories of patients. Antimicrobial Agents and Chemotherapy. 2011;**55**(7):3284-3294. DOI: 10.1128/AAC.01733-10

[77] Karaiskos I, Friberg LE, Pontikis K, Ioannidis K, Tsagkari V, Galani L, et al. Colistin population pharmacokinetics after application of a loading dose of 9 MU colistin methanesulfonate in critically ill patients. Antimicrobial Agents and Chemotherapy. 2015;**59**(12):7240-7248. DOI: 10.112 8/AAC.00554-15

[78] Sandri AM, Landersdorfer CB, Jacob J, Boniatti MM, Dalarosa MG, Falci DR, et al. Population pharmacokinetics of intravenous polymyxin B in critically ill patients: Implications for selection of dosage regimens. Clinical Infectious Diseases. 2013;**57**(4):524-531. DOI: 10.1093/cid/cit 334

[79] Falagas ME, Kasiakou SK. Colistin: The revival of polymyxins for the management of multidrug-resistant gram-negative bacterial infections. Clinical Infectious Diseases. 2005;**40**(9):1333-1341. DOI: 10.1086/429323

[80] Fass RJ, Barnishan J, Ayers LW. Emergence of bacterial resistance to imipenem and ciprofloxacin in a university hospital. The Journal of Antimicrobial Chemotherapy. 1995;**36**(2):343-353. DOI: 10.1093/jac/36.2.343

[81] Levin AS, Mendes CM, Sinto SI, Sader HS, Scarpitta CR, Rodrigues E, et al. An outbreak of multiresistant *Acinetobacter baumanii* in a university hospital in São Paulo, Brazil. Infection Control and Hospital Epidemiology. 1996;**17**(6):366-368. DOI: 10.1086/647319

[82] Dalla-Costa LM, Coelho JM, Souza HA, Castro ME, Stier CJ, Bragagnolo KL, et al. Outbreak of carbapenem-resistant *Acinetobacter baumannii* producing the OXA-23 enzyme in Curitiba, Brazil. Journal of Clinical Microbiology. 2003;**41**(7):3403-3406. DOI: 10.1128/JCM.41.7.3403-34 06.2003

[83] Evans ME, Feola DJ, Rapp RP. Polymyxin B sulfate and colistin: Old antibiotics for emerging multiresistant gram-negative bacteria. The Annals of Pharmacotherapy. 1999;**33**(9):960-967. DOI: 10.1345/aph.18426

[84] Falagas ME, Kasiakou SK. Toxicity of polymyxins: A systematic review of the evidence from old and recent studies. Critical Care. 2006;**10**(1):R27. DOI: 10.11 86/cc3995

[85] Kunin CM, Bugg A. Binding of polymyxin antibiotics to tissues: The major determinant of distribution and persistence in the body. The Journal of Infectious Diseases. 1971;**124**(4):394-400. DOI: 10.1093/infdis/124.4.394

[86] Koch-Weser J, Sidel VW, Federman EB, Kanarek P, Finer DC, Eaton AE. Adverse effects of sodium colistimethate. Manifestations and specific reaction rates during 317 courses of therapy. Annals of Internal Medicine. 1970;**72**(6):857-868. DOI: 10.7326/0003-4819-72-6-857

[87] Ledson MJ, Gallagher MJ, CowperthwaiteC, ConveryRP, WalshawMJ. Four years experience of intravenous colomycin in an adult cystic fibrosis unit. The European Respiratory Journal. 1998;**12**(3):592-594. DOI: 10.1183/09031936.98.12030592

[88] Price DJ, Graham DI. Effects of large doses of colistin sulphomethate sodium on renal function. British Medical Journal. 1970;**4**(5734):525-527. DOI: 10.1136/bmj.4.5734.5 25

[89] Tallgren LG, Liewendahl K, Kuhlbaeck B. The therapeutic success and nephrotoxicity of colistin in acute and chronic nephropathies with impaired renal function. Acta Medica Scandinavica. 1965;**177**:717-728. DOI: 10.1111/j.0954-6820.1965. tb01882.x

[90] Abdelraouf K, He J, Ledesma KR, Hu M, Tam VH. Pharmacokinetics and renal disposition of polymyxin B in an animal model. Antimicrobial Agents and Chemotherapy. 2012;**56**(11):5724-5727. DOI: 10.1128/AAC.01333-12

[91] Akajagbor DS, Wilson SL, Shere-Wolfe KD, Dakum P, Charurat ME, Gilliam BL. Higher incidence of acute kidney injury with intravenous colistimethate sodium compared with polymyxin B in critically ill patients at a tertiary care medical center. Clinical Infectious Diseases. 2013;**57**(9):1300-1303. DOI: 10.1093/cid/cit453

[92] Phe K, Lee Y, McDaneld PM, Prasad N, Yin T, Figueroa DA, et al. In vitro assessment and multicenter cohort study of comparative nephrotoxicity rates associated with colistimethate versus polymyxin B therapy. Antimicrobial Agents and Chemotherapy. 2014;**58**(5):2740-2746. DOI: 10.1128/AAC.024 76-13

[93] Nation RL, Rigatto MHP, Falci DR, Zavascki AP. Polymyxin acute kidney injury: Dosing and other strategies to reduce toxicity. Antibiotics (Basel). 2019;**8**(1):24. DOI: 10.3390/antibiotics8010024

[94] Zavascki AP, Nation RL. Nephrotoxicity of Polymyxins: Is there any difference between colistimethate and polymyxin B? Antimicrobial Agents and Chemotherapy. 2017;**61**(3):e02319-e02316. DOI: 10.1128/AAC.02319-16

[95] Deng Y, Gu JY, Li X, Tong H, Guo SW, Xu B, et al. Does monitoring total and free polymyxin b1 plasma concentrations predict polymyxin B-induced nephrotoxicity? A retrospective study in critically ill patients. Infectious Disease and Therapy. 2022;**11**(4):1591-1608. DOI: 10.1007/s40121-022-00655-3

[96] Li J, Nation RL, Milne RW, Turnidge JD, Coulthard K. Evaluation of colistin as an agent against multiresistant gram-negative bacteria. International Journal of Antimicrobial Agents. 2005;**25**(1):11-25. DOI: 10.1016/j.ijantimicag.2004.10.001

[97] Pastewski AA, Caruso P, Parris AR, Dizon R, Kopec R, Sharma S, et al. Parenteral polymyxin B use in patients with multidrug-resistant gram-negative bacteremia and urinary tract infections: A retrospective case series. The Annals of Pharmacotherapy. 2008;**42**(9):1177-1187. DOI: 10.1345/aph.1K346

[98] Elias LS, Konzen D, Krebs JM, Zavascki AP. The impact of polymyxin B dosage on in-hospital mortality of patients treated with this antibiotic. The Journal of Antimicrobial Chemotherapy. 2010;**65**(10):2231-2237. DOI: 10.1093/jac/dkq285

[99] Kubin CJ, Ellman TM, Phadke V, Haynes LJ, Calfee DP, Yin MT. Incidence and predictors of acute kidney injury associated with intravenous polymyxin B therapy. The Journal of Infection. 2012;**65**(1):80-87. DOI: 10.1016/j.jinf.2012.01.015

[100] Mingeot-Leclercq MP, Tulkens PM, Denamur S, Vaara T, Vaara M. Novel

polymyxin derivatives are less cytotoxic than polymyxin B to renal proximal tubular cells. Peptides. 2012;**35**(2):248-252. DOI: 10.1016/j.peptides.2012.03. 033

[101] Falagas ME, Kyriakidou M, Voulgaris GL, Vokos F, Politi S, Kechagias KS. Clinical use of intravenous polymyxin B for the treatment of patients with multidrug-resistant gram-negative bacterial infections: An evaluation of the current evidence. The Journal of Global Antimicrobial Resistance. 2021;**24**:342-359. DOI: 10.1016/j.jgar.202 0.12.026

[102] Aggarwal R, Dewan A. Comparison of nephrotoxicity of colistin with polymyxin B administered in currently recommended doses: A prospective study. Annals of Clinical Microbiology and Antimicrobials. 2018;**17**(1):15. DOI: 10.1186/s12941-018-0262-0

[103] Levin AS, Barone AA, Penço J, Santos MV, Marinho IS, Arruda EA, et al. Intravenous colistin as therapy for nosocomial infections caused by multidrug-resistant *Pseudomonas aeruginosa* and *Acinetobacter baumannii*. Clinical Infectious Diseases. 1999;**28**(5):1008-1011. DOI: 10.1086/514732

[104] Garnacho-Montero J, Ortiz-Leyba C, Jiménez-JiménezFJ,Barrero-AlmodóvarAE, García-Garmendia JL, Bernabeu-WittelI M, et al. Treatment of multidrug-resistant *Acinetobacter baumannii* ventilator-associated pneumonia (VAP) with intravenous colistin: A comparison with imipenem-susceptible VAP. Clinical Infectious Diseases. 2003;**36**(9):1111-1118. DOI: 10.1086/374337

[105] Ouderkirk JP, Nord JA, Turett GS, Kislak JW. Polymyxin B nephrotoxicity and efficacy against nosocomial infections caused by multiresistant gram-negative bacteria. Antimicrobial Agents and Chemotherapy. 2003;**47**(8):2659-2662. DOI: 10.1128/AAC.47.8.2659-2662.2003

[106] Markou N, Apostolakos H, KoumoudiouC,AthanasiouM,KoutsoukouA, Alamanos I, et al. Intravenous colistin in the treatment of sepsis from multiresistant gram-negative bacilli in critically ill patients. Critical Care. 2003;7(5):R78-R83. DOI: 10.1186/cc2358

[107] Sobieszczyk ME, Furuya EY, Hay CM, Pancholi P, Della-Latta P, Hammer SM, et al. Combination therapy with polymyxin B for the treatment of multidrug-resistant gram-negative respiratory tract infections. The Journal of Antimicrobial Chemotherapy. 2004;**54**(2):566-569. DOI: 10.1093/jac/dkh369

[108] Falagas ME, Rizos M, Bliziotis IA, Rellos K, Kasiakou SK, Michalopoulos A. Toxicity after prolonged (more than four weeks) administration of intravenous colistin. BMC Infectious Diseases. 2005;**5**:1. DOI: 10.1186/1471-2334-5-1

[109] Reina R, Estenssoro E, Sáenz G, Canales HS, Gonzalvo R, Vidal G, et al. Safety and efficacy of colistin in *Acinetobacter* and *Pseudomonas* infections: A prospective cohort study. Intensive Care Medicine. 2005;**31**(8):1058-1065. DOI: 10.1007/s00134-005-2691-4

[110] Michalopoulos AS, Tsiodras S, Rellos K, Mentzelopoulos S, Falagas ME. Colistin treatment in patients with ICU-acquired infections caused by multiresistant gram-negative bacteria: The renaissance of an old antibiotic. Clinical Microbiology and Infection. 2005;**11**(2):115-121. DOI: 10.1111/j.1469-0691.2004.01043.x

[111] Kasiakou SK, Michalopoulos A, Soteriades ES, Samonis G, Sermaides GJ, Falagas ME. Combination therapy with

intravenous colistin for management of infections due to multidrug-resistant gram-negative bacteria in patients without cystic fibrosis. Antimicrobial Agents and Chemotherapy. 2005;**49**(8):3136-3146. DOI: 10.1128/AAC.49.8.3136-3146.2005

[112] Holloway KP, Rouphael NG, Wells JB, King MD, Blumberg HM. Polymyxin B and doxycycline use in patients with multidrug-resistant *Acinetobacter baumannii* infections in the intensive care unit. The Annals of Pharmacotherapy. 2006;**40**(11):1939-1945. DOI: 10.1345/aph.1H353

[113] Ramasubban S, Majumdar A, Das PS. Safety and efficacy of polymyxin B in multidrug-resistant gram-negative severe sepsis and septic shock. Indian Journal of Critical Care Medicine. 2008;**12**(4):153-157. DOI: 10.4103/0972-5229.45074

[114] Oliveira MS, Prado GV, Costa SF, Grinbaum RS, Levin AS. Polymyxin B and colistimethate are comparable as to efficacy and renal toxicity. Diagnostic Microbiology and Infectious Disease. 2009;**65**(4):431-434. DOI: 10.1016/j.diagmicrobio.2009.07.018

[115] Mendes CA, Cordeiro JA, Burdmann EA. Prevalence and risk factors for acute kidney injury associated with parenteral polymyxin B use. The Annals of Pharmacotherapy. 2009;**43**(12):1948-1955. DOI: 10.1345/aph.1M277

[116] Esaian D, Dubrovskaya Y, Phillips M, Papadopoulos J. Effectiveness and tolerability of a polymyxin B dosing protocol. The Annals of Pharmacotherapy. 2012;**46**(3):455-456. DOI: 10.1345/aph.1Q294

[117] Nandha R, Sekhri K, Mandal AK. To study the clinical efficacy and nephrotoxicity along with the risk factors for acute kidney injury associated with parenteral polymyxin B. Indian Journal of Critical Care Medicine. 2013;**17**(5):283-287. DOI: 10.4103/0972-5229.120319

[118] Crusio R, Rao S, Changawala N, Paul V, Tiu C, van Ginkel J, et al. Epidemiology and outcome of infections with carbapenem-resistant gram-negative bacteria treated with polymyxin B-based combination therapy. Scandinavian Journal of Infectious Diseases. 2014;**46**(1):1-8. DOI: 10.3109/00365548.2013.844350

[119] Tuon FF, Rigatto MH, Lopes CK, Kamei LK, Rocha JL, Zavascki AP. Risk factors for acute kidney injury in patients treated with polymyxin B or colistin methanesulfonate sodium. International Journal of Antimicrobial Agents. 2014;**43**(4):349-352. DOI: 10.1016/j.ijantimicag.2013.12.002

[120] Siddiqui NU, Qamar FN, Jurair H, Haque A. Multidrug-resistant gram-negative infections and use of intravenous polymyxin B in critically ill children of a developing country: Retrospective cohort study. BMC Infectious Diseases. 2014;**14**:626. DOI: 10.1186/s12879-014-0626-9

[121] Rigatto MH, Behle TF, Falci DR, Freitas T, Lopes NT, Nunes M, et al. Risk factors for acute kidney injury (AKI) in patients treated with polymyxin B and influence of AKI on mortality: A multicentre prospective cohort study. The Journal of Antimicrobial Chemotherapy. 2015;**70**(5):1552-1557. DOI: 10.1093/jac/dku561

[122] Nelson BC, Eiras DP, Gomez-Simmonds A, Loo AS, Satlin MJ. Jenkins SGet al. Clinical outcomes associated with polymyxin B dose in patients with bloodstream infections

due to carbapenem-resistant gram-negative rods. Antimicrobial Agents and Chemotherapy. 2015;**59**(11):7000-7006. DOI: 10.1128/AAC.00844-15

[123] Dubrovskaya Y, Prasad N, Lee Y, Esaian D, Figueroa DA, Tam VH. Risk factors for nephrotoxicity onset associated with polymyxin B therapy. The Journal of Antimicrobial Chemotherapy. 2015;**70**(6):1903-1907. DOI: 10.1093/jac/dkv014

[124] Rigatto MH, Oliveira MS, Perdigão-Neto LV, Levin AS, Carrilho CM, Tanita MT, et al. Multicenter prospective cohort study of renal failure in patients treated with colistin versus polymyxin B. Antimicrobial Agents and Chemotherapy. 2016;**60**(4):2443-2449. DOI: 10.1128/AAC. 02634-15

[125] Cai B, Cai Y, Liew YX, Chua NG, Teo JQ, Lim TP, et al. Clinical efficacy of polymyxin monotherapy versus nonvalidated polymyxin combination therapy versus validated polymyxin combination therapy in extensively drug-resistant gram-negative bacillus infections. Antimicrobial Agents and Chemotherapy. 2016;**60**(7):4013-4022. DOI: 10.1128/AAC.03 064-15

[126] Ismail B, Shafei MN, Harun A, Ali S, Omar M, Deris ZZ. Predictors of polymyxin B treatment failure in gram-negative healthcare-associated infections among critically ill patients. Journal of Microbiology, Immunology, and Infection. 2018;**51**(6):763-769. DOI: 10.1016/j.jmii.2017.03.007

[127] John JF, Falci DR, Rigatto MH, Oliveira RD, Kremer TG, Zavascki AP. Severe infusion-related adverse events and renal failure in patients receiving high-dose intravenous polymyxin B. Antimicrobial Agents and Chemotherapy. 2017;**62**(1):e01617-e01617. DOI: 10.1128/AAC.01617-17

[128] Okoduwa A, Ahmed N, Guo Y, Scipione MR, Papadopoulos J, Eiras DP, et al. Nephrotoxicity associated with intravenous polymyxin B once-versus twice-daily dosing regimen. Antimicrobial Agents and Chemotherapy. 2018;**62**(8):e00025-e00018. DOI: 10.1128/AAC.00025-18

[129] Zheng JY, Huang SS, Huang SH, Ye JJ. Colistin for pneumonia involving multidrug-resistant *Acinetobacter calcoaceticus-Acinetobacter baumannii* complex. Journal of Microbiology, Immunology, and Infection. 2020;**53**(6):854-865. DOI: 10.1016/j. jmii.2019. 08.007

[130] Li KL, Abad CLR. The clinical profile and outcomes of adult patients given intravenous colistin for multidrug-resistant gram-negative infections in a Philippine tertiary hospital. International Journal of Infectious Diseases. 2020;**93**:9-14. DOI: 10.1016/j.ijid. 2020.01.022

[131] Kagami K, Ishiguro N, Yamada T, Niinuma Y, Iwasaki S, Taki K, et al. Efficacy and safety of colistin for the treatment of infections caused by multidrug-resistant gram-negative bacilli. Journal of Infection and Chemotherapy. 2021;**27**(3):473-479. DOI: 10.1016/j.jiac.2020.10.024

[132] Aitullina A, Purviņa S, Krūmiņa A. Colistin co-administration with other nephrotoxins: Experience of teaching hospital of Latvia. International Journal of Clinical Pharmacy. 2021;**43**(3):509-517. DOI: 10.1007/s11096-020-01154-6

[133] Saelim W, Changpradub D, ThunyaharnS,JuntanawiwatP,NulsopaponP, Santimaleeworagun W. Colistin plus sulbactam or fosfomycin against carbape

nem-resistant *Acinetobacter baumannii*: Improved efficacy or decreased risk of nephrotoxicity? Infection & Chemotherapy. 2021;**53**(1):128-140. DOI: 10.3947/ic.202 1.0007

[134] Jeong YJ, Gu N, Kwack WG, Kang Y, Park SY, Yoon YS. Prospective observational study of the impact of plasma colistin levels in patients with carbapenem-resistant *Acinetobacter baumannii* pneumonia. The Journal of Global Antimicrobial Resistance. 2021;**27**:315-323. DOI: 10.1016/j.jgar.2021.10.017

[135] Sadyrbaeva-Dolgova S, García-Fumero R, Exposito-Ruiz M, Pasquau-Liaño J, Jiménez-Morales A, Hidalgo-Tenorio C. Incidence of nephrotoxicity associated with intravenous colistimethate sodium administration for the treatment of multidrug-resistant gram-negative bacterial infections. Scientific Reports. 2022;**12**(1):15261. DOI: 10.1038/s41598-022-19626-2

[136] Keski NAS, Seyman D, Önder KD, Kizilateş F, Keski NO. Investigation of effect of the colistin loading dosage on the clinical, microbiological, and laboratory results in *Acinetobacter baumannii* ventilator-associated pneumonia/pneumonia. International Journal of Clinical Practice. 2022;**2022**:5437850. DOI: 10.1155/2022/5437850

[137] Koza Y. Acute kidney injury: Current concepts and new insights. Journal of Injury and Violence Research. 2016;**8**(1):58-62. DOI: 10.5249/jivr.v8i1.610

[138] Wasung ME, Chawla LS, Madero M. Biomarkers of renal function, which and when? Clinica Chimica Acta. 2015;**438**:350-357. DOI: 10.1016/j.cca.2014.08.039

[139] Wolinsky E, Hines JD. Neurotoxic and nephrotoxic effects of colistin in patients with renal disease. The New England Journal of Medicine. 1962;**266**:759-762. DOI: 10.1056/NEJM196204122661505

[140] Lindesmith LA, Baines RD Jr, Bigelow DB, Petty TL. Reversible respiratory paralysis associated with polymyxin therapy. Annals of Internal Medicine. 1968;**68**(2):318-327. DOI: 10.7326/0003-4819-68-2-318

[141] Cox CE, Harrison LH. Intravenous sodium colistimethate therapy of urinary tract infections: Pharmacological and bacteriological studies. Antimicrobial Agents and Chemotherapy (Bethesda). 1970;**10**:296-302

[142] Fekety FR Jr, Norman PS, Cluff LE. The treatment of gram-negative bacillary infections with colistin. The toxicity and efficacy of large doses in forty-eight patients. Annals of Internal Medicine. 1962;**57**:214-229. DOI: 10.7326/0003-4819-57-2-214

[143] Hopper J Jr, Jawetz E, Hinman F Jr. Polymyxin B in chronic pyelonephritis: Observations on the safety of the drug and on its influence on the renal infection. The American Journal of the Medical Sciences. 1953;**225**(4):402-409

[144] Mcmillan M, Price TM, Maclaren DM, Scott GW. *Pseudomonas pyocyanea* infection treated with colistin methane sulphonate. Lancet. 1962;**2**(7259):737-739. DOI: 10.1016/s0140-6736(62)90568-8

[145] Dai C, Ciccotosto GD, Cappai R, Tang S, Li D, Xie S, et al. Curcumin attenuates colistin-induced neurotoxicity in N2a cells via anti-inflammatory activity, suppression of oxidative stress, and apoptosis. Molecular Neurobiology. 2018;**55**(1):421-434. DOI: 10.1007/s12035-016-0276-6

[146] Dai C, Li J, Lin W, Li G, Sun M, Wang F, et al. Electrophysiology and ultrastructural changes in mouse sciatic nerve associated with colistin sulfate exposure. Toxicology Mechanisms and Methods. 2012;**22**(8):592-596. DOI: 10.3109/15376516.2012.704956

[147] Dai C, Li J, Li J. New insight in colistin induced neurotoxicity with the mitochondrial dysfunction in mice central nervous tissues. Experimental and Toxicologic Pathology. 2013;**65**(6):941-948. DOI: 10.1016/j.etp.2013.01.008

[148] Dai C, Tang S, Velkov T, Xiao X. Colistin-induced apoptosis of neuroblastoma-2a cells involves the generation of reactive oxygen species, mitochondrial dysfunction, and autophagy. Molecular Neurobiology. 2016;**53**(7):4685-4700. DOI: 10.10 07/s12035-015-9396-7

[149] Dai C, Ciccotosto GD, Cappai R, Wang Y, Tang S, Xiao X, et al. Minocycline attenuates colistin-induced neurotoxicity via suppression of apoptosis, mitochondrial dysfunction, and oxidative stress. The Journal of Antimicrobial Chemotherapy. 2017;**72**(6):1635-1645. DOI: 10.1093/jac/dkx037

[150] Mattos KPH, Cintra ML, Gouvêa IR, Ferreira LÁ, Velho PENF, Moriel P. Skin hyperpigmentation following intravenous polymyxin B treatment associated with melanocyte activation and inflammatory process. Journal of Clinical Pharmacy and Therapeutics. 2017;**42**(5):573-578. DOI: 10.1111/jcpt.12543

[151] Knueppel RC, Rahimian J. Diffuse cutaneous hyperpigmentation due to tigecycline or polymyxin B. Clinical Infectious Diseases. 2007;**45**(1):136-138. DOI: 10.1086/518706

[152] Mattos KP, Lloret GR, Cintra ML, Gouvêa IR, Betoni TR, Mazzola PG, et al. Acquired skin hyperpigmentation following intravenous polymyxin B treatment: A cohort study. Pigment Cell & Melanoma Research. 2016;**29**(3):388-390. DOI: 10.11 11/pcmr.12468

[153] Gothwal S, Meena K, Sharma SD. Polymyxin B induced generalized hyperpigmentation in neonates. Indian Journal of Pediatrics. 2016;**83**(2):179-180. DOI: 10.1007/s12098-015-1798-z

[154] Zavascki AP, Manfro RC, Maciel RA, Falci DR. Head, and neck hyperpigmentation probably associated with polymyxin B therapy. The Annals of Pharmacotherapy. 2015;**49**(10):1171-1172. DOI: 10.1177/1060028015595643

[155] Lahiry S, Choudhury S, Mukherjee A, Bhunya PK, Bala M. Polymyxin B-induced diffuse cutaneous hyperpigmentation. Journal of Clinical and Diagnostic Research. 2017;**11**(2):FD01-FD02. DOI: 10.7860/JCDR/2017/24278.9213

[156] Zavascki AP, Schuster LF, Duquia RP. Histopathological findings of pigmented lesion and recovery of natural skin color in a patient with polymyxin B-associated diffuse hyperpigmentation. International Journal of Antimicrobial Agents. 2016;**48**(5):579-580. DOI: 10.1016/j.ijantimicag.2016.08.010

[157] Zheng G, Cao L, Che Z, Mao E, Chen E, He J. Polymyxin B-induced skin hyperpig mentation: A rare case report and literature review. BMC Pharmacology and Toxicology. 2018;**19**(1):41. DOI: 10.1186/s40360-018-0226-1

[158] Li YM, Milikowski C, Selvaggi G, Abbo LM, Skiada D, Galimberti F. Polymyxin B-induced skin hyperpigmentation. Transplant

Infectious Disease. 2020;**22**(5):e13312. DOI: 10.1111/tid.13312

[159] Silpa-Archa N, Kohli I, Chaowattanapanit S, Lim HW, Hamzavi I. Postinflammatory hyperpigmentation: A comprehensive overview: Epidemiology, pathogenesis, clinical presentation, and noninvasive assessment technique. Journal of the American Academy of Dermatology. 2017;**77**(4):591-605. DOI: 10.1016/j.jaad.2017.01.035

[160] Lu C, Hou N. Skin hyperpigmentation in coronavirus disease 2019 patients: Is polymyxin B the culprit? Frontiers in Pharmacology. 2020;**11**:01304. DOI: 10.3389/fphar.2020.01304

Section 4

Acinetobacter baumannii in Latin America

Chapter 7

Carbapenem-Resistant *Acinetobacter baumannii* in Latin America

Thiago Pavoni Gomes Chagas, Karyne Rangel and Salvatore Giovanni De-Simone

Abstract

Acinetobacter baumannii is an important bacterial pathogen associated with healthcare-associated infections (HAIs), especially in critically ill patients admitted to Intensive Care Units (ICU). Its ability to acquire antibiotic resistance determinants has propelled its clinical relevance. The rise in *Acinetobacter* infections and hospital outbreaks have been extensively described worldwide and are usually caused by carbapenem-resistant isolates. To compound the problem, Carbapenem-resistant A. *baumannii* (CRAb) isolates are also resistant to a wide range of other antibiotics, representing a serious threat to public health. Since 2017, *A. baumannii* has been listed as a critical priority pathogen that poses a great threat to human health, according to the World Health Organization (WHO). The carbapenem-resistant rates in *A. baumannii* are notorious around the world. However, Latin America has one of the highest in the world. Carbapenem resistance in *A. baumannii* is due mainly to the presence of horizontally acquired OXA-type carbapenem resistance genes, including $bla_{\text{OXA-23}}$, in most regions. Thus, this review aims to summarize the distribution of CRAb and its major carbapenem resistance mechanisms in Latin America.

Keywords: *Acinetobacter baumannii*, carbapenems, antimicrobial resistance, carbapenemases, oxacillinases

1. Introduction

Acinetobacter baumannii is Gram-negative, nonfermenting, aerobic coccobacilli, catalase-positive, oxidase-negative, and non-motile [1, 2]. It has also been considered the most serious among the 'ESKAPE' (*Enterococcus faecium*, *Staphylococcus aureus*, *Klebsiella pneumoniae*, *A. baumannii*, *Pseudomonas aeruginosa*, and *Enterobacter* species), a group of six pathogens with multidrug resistance and virulence factors [3].

In the *Acinetobacter* genus, *A.baumannii* is a more relevant species grouped as the *Acinetobacter calcoaceticus–Acinetobacter baumannii* (ACB) complex [4]. Clinical samples frequently recover this microorganism. It has been responsible for many nosocomial infection outbreaks in Intensive Care Units (ICU) [5]. *A. baumannii* can also be associated with community-acquired infections such as pneumonia and bacteremia. However, these community infections are less common and have been associated with comorbidities (e.g., alcoholism, smoking, diabetes mellitus, chronic obstructive pulmonary disease, and renal disease) [6, 7].

A. baumannii has been recognized as causing severe healthcare-associated infections (HAIs) [6]. This Gram-negative pathogen has been associated with pneumonia, endocarditis, bacteremia, wound infections, urinary tract infections, and meningitis in hospital settings. However, ventilator-associated pneumonia and bloodstream infections are the most important infections, accompanied by the highest mortality rates [3, 5, 6, 8]. Risk factors associated with colonization or infection include intensive care unit admission, invasive medical procedures, prolonged hospitalization, antimicrobial agent exposure, prior hospitalization, and local colonization pressure on susceptible patients [8].

The ability to resist the vast majority of available antimicrobial agents is an important determinant in clinical outcomes of *A. baumannii* infections and spread in the hospital setting [7, 9, 10]. Multidrug-resistant isolates of *A. baumannii* have been reported increasingly during the last decade [11, 12]. Previous studies indicated that the estimated global incidence of *A. baumannii* infections is approximately 1,000,000 cases annually, of which 50% are resistant to multiple antibiotics, including carbapenems [13, 14]. Carbapenem-resistant *A. baumannii* (CRAb) isolates have been increasingly observed worldwide, constituting a serious threat to public health [12], especially in Latin America [15], being significantly associated with increased mortality.

2. Carbapenem-resistance *A. baumannii*

Antimicrobial resistance (AMR) has emerged as one of the global healthcare threats of the twenty-first century [16]. Projections estimated 10 million deaths per year attributable to bacterial AMR by 2050 [17, 18]. *A. baumannii* strains can develop resistance to all the antibiotics available, and outbreaks caused by multidrug-resistant (MDR), extensively drug-resistant (XDR) and even pan-drug-resistant (PDR) strains have been reported around the world [19].

Different global health authorities, including the European Centre for Disease Prevention and Control (ECDC), Infectious Diseases Society of America (IDSA), and Center for Disease Control and Prevention (CDC) have appointed MDR *A. baumannii* a critical threat to global health [20–23]. 2017, the World Health Organization (WHO) listed CRAb as a crucial priority due to its high AMR rates [24]. The rise of CRAb strains as an opportunistic pathogen poses a significant threat to global health.

2.1 Carbapenems

Carbapenems, such as the most popular imipenem and meropenem, play a critically important role as a therapeutic option for serious infections caused by MDR *A. baumannii* [8] due to their effective activity and their safety [25, 26]. This β-lactam subclass demonstrates a wider range of antimicrobial activity than penicillins, cephalosporins, or β-lactam/ β-lactamase inhibitor combinations [27].

Generally, they have excellent bactericidal activity and stability toward a range of β-lactamases, except the emerging carbapenemases [8, 28, 29]. Carbapenems (except ertapenem that is inactive against *Pseudomonas* and *A. baumannii*) displayed activity against both Gram-negatives (except *Stenotrophomonas maltophilia*) and Gram-positive bacteria (except methicillin-resistant *S. aureus*, *E. faecium* and *Enterococcus fecalis* apart from imipenem) [30].

Like other β-lactams, carbapenems are bactericidal agents that bind to the penicillin-binding proteins (PBPs), inhibiting bacterial cell wall synthesis [31]. Specifically, they prevent transpeptidation [32]. Conventionally, that β-lactam class enters Gram-negative bacteria through outer membrane proteins (OMPs), also known as porins [28].

A classification system for carbapenems was proposed based on their antimicrobial activity, dividing them into three groups. Carbapenems group 1, which included ertapenem, are inefficient against non-fermentative Gram-negative bacilli and may be more suitable for community-acquired infections. Carbapenems from group 2, such as meropenem, imipenem, and doripenem, have broad-spectrum actions, are active against non-fermentative Gram-negative bacilli, and are effective against nosocomial infections. Group 3 carbapenems are potent against non-fermentative Gram-negative bacilli and *S. aureus*, which are resistant to methicillin [33–35].

Carbapenems have low oral bioavailability and must be administered intravenously because they cannot cross the gastrointestinal membranes readily. Additionally, imipenem-cilastatin and ertapenem can also be administered intramuscularly. All these carbapenem antibiotics are excreted via the kidneys [28].

These agents have a role as empirical and definitive therapy options in a range of serious infections. In ICU, carbapenems are especially valuable in units with known third-generation cephalosporin resistance problems, in patients with disease who have received previous antibiotic courses, and in polymicrobial infections [36]. Carbapenems are appropriate for use in the lower respiratory tract, skin and soft tissue, central nervous system, urinary tract, joint, muscle, gynecologic, obstetric, and abdominal infections or in the management of febrile neutropenia and problems due to cystic fibrosis [27].

Since the first CRAb was identified in 1991, there has been a considerable increase in the amount of *A. baumannii* strains that have acquired resistance to this β-lactam class [37]. This problem is critical, especially considering that most CRAb strains resist other antibiotic classes.

2.2 Treatment options

When carbapenem resistance is suspected and/or determined, some agents can be used in therapeutic combinations to treat CRAb infections, for example, β-lactamase inhibitors such as sulbactam; polymyxins, tetracyclines, such as minocycline and doxycycline; fosfomycin, rifamycin, and carbapenem therapy combined with other antibiotics [38].

2.3 Global rates of CRAb

Carbapenem resistance rates can vary according to the geographic Region around the world. Among 2.674 *A. baumannii* isolates collected from 13 countries in the Asia-Pacific region by Antimicrobial Testing Leadership and Surveillance (ATLAS) program between 2012 and 2019, carbapenem resistance rates ranged from the lowest in Japan (2.8%) and Australia (6.5%) to the highest in South Korea (88%). According to the previous review, CRAb is critically problematic across Asia and the Americas, except in Japan (3.5%) and Canada (4.7%). Oceania, Western Europe, the Nordic Region, and part of central Europe have the lowest rates (<10%). However, in areas surrounding the Mediterranean, including southern Europe, the Middle East, and North Africa, up to 90% of strains are resistant to carbapenems [39].

The latest Surveillance of AMR in Europe 2022 reports the total carbapenem resistance rate ranged from 31.9 to 38% among *Acinetobacter* spp. Isolates from 2016 to 2020. The percentages of carbapenem-resistant *Acinetobacter* spp. Varied within the Region in 2020, from below 1% in three (8%) of 38 countries/areas (Ireland, the Netherlands, and Norway) to percentages equal to or above 50% in 21 (55%) countries/areas, mostly in Southern and Eastern Europe [40]. The number of European countries with 50% or higher carbapenem resistance rates increased from 12 in 2015–2018 to 21 countries in 2018–2020 [40, 41].

From 2012 to 2017, the incidence of CRAb from clinical cultures decreased in the United States. The Centers for Disease Control and Prevention (CDC) estimated 8500 cases among U.S. hospitalized patients in 2017, resulting in 700 deaths [42]. Between 2013 and 2016, the SENTRY Antimicrobial Surveillance Program reported, among ACB complex Isolates, a susceptibility rate for meropenem of 54.9% in North America [43]. For imipenem, the susceptibility rate was 57.7%. Comparing the intervals 1997–2000 and 2013–2016, the susceptibility rate for meropenem significantly decreased from 88.8 to 54.9% [43].

Among 4.320 *A. baumannii* isolates collected across different regions of the world between 2016 and 2018 by Seifert et al. [44], the global resistance rate for meropenem was 64.4%. The highest meropenem resistance rates observed were in Africa/ Middle East (81.1%), Latin America (78.4%), Asian/South Pacific (67.5%), and Europe (63%) [44].

2.4 CRAb in Latin America

Rates of carbapenem resistance among *A. baumannii* in Latin America appear to be one of the highest in the world. These rates up to 90% for *A. baumannii* isolates can be found across the different countries of Latin America, with the resistance rate of *A. baumannii* isolates greater than 50% in many countries [15]. In a review by Ma and McClean [39], Carbapenem resistance rates ranged from 0 to 97.5% among Latin American isolates [39].

ACB complex Isolates were collected from 17 Latin America centers (7 countries) from January 1997 to December 2016 through the SENTRY Program. Data of this Surveillance program appointed a susceptibility rate for meropenem of 13.7%. For imipenem, this resistance rate was 14.4%. The susceptibility rates declined continuously in Latin America's 2009–2012 and 2013–2016 periods [43].

The global dissemination of CRAb is associated with clonal lineages, illustrating this organism's success in acquiring carbapenem resistance [45]. Initially, three disseminated lineages of *A. baumannii* called European clones I, II, and III were characterized in European countries. Posteriorly, complementary studies showed that these lineages had already spread worldwide, and thus, European clones were renamed international clonal (IC) lines I, II, and III [7, 45, 46]. At the moment, molecular epidemiological studies have recognized nine major International Clones (1–9) of *A. baumannii*, the most widespread of which is IC 2 (II) [47]. However, CRAb isolates in Latin America are not associated with the most pervasive IC2 [48, 49].

In Latin countries such as Brazil, Argentina, Chile, and Paraguay, the major CRAb clones were found to belong to IC 4 and IC 5 [49, 50]. These IC 4 and IC 5 correspond to clonal complexes $CC15^{Past}$/$CC103^{OXF}$ and $CC79^{Past}$/$CC227^{OXF}$ defined by Pasteur (^{Past}) and Oxford (^{OXF}) Multilocus Sequence Type (MLST) schemes [49–51]. Other ICs have been observed in Latin regions, such as IC 1 ($CC1^{Past}$/ $CC109^{OXF}$), IC 2 ($CC2^{Past}$/$CC92^{OXF}$), IC 6 ($CC78^{Past}$/$CC944^{OXF}$) and IC7 ($CC25^{Past}$/ $CC110^{OXF}$) [50, 52–55].

2.5 CRAb in COVID-19 pandemic

Latin America has faced critical moments during the COVID-19 pandemic and was considered one of the world epicenters [56]. Hospitalization of COVID-19 patients predisposed to severe consequences such as HAIs and secondary or coinfections associated with MDR bacteria such as *A. baumannii* [57–59]. Since the beginning of COVID-19, the emergence of resistant microorganisms causing HAIs has been documented [60].

An increased risk of CRAb infections in patients with an increased risk of mortality due to COVID-19 infections was reported. This increased incidences of *A. baumannii* infections during the COVID-19 pandemic were related to various reasons such as prolonged hospital stay, mechanical ventilation, and immunosuppression [61].

In a retrospective analysis of two prospective observational cohort studies of COVID-19 patients in 10 countries, including Colombia, Chile, Ecuador, Mexico, Argentina, Uruguay, and Brazil, *A. baumannii* was Latin America's fourth most prevalent bacteria (10.6%). However, this bacteria was less predominant in Europe [62].

A recent study reported the occurrence of CRAb belonging to IC 2 caused a large outbreak among COVID-19 patients at a public hospital in Brazil [63]. At an Argentinian hospital, an experience with carbapenem-resistant isolates such as CRAb during the period with active cases of COVID-19 was reported [64]. Loyola-Cruz et al. described *A. baumannii* involved in outbreaks non-detected in COVID-19 patients at a Mexican hospital. Among 14 *A. baumannii* isolates, meropenem and imipenem resistance rates were 100% [65]. Another Mexican study conducted by Alcántar-Curiel et al. [66] reported 34% (n = 39) of CRAb isolates linked to nosocomial bacteremias in COVID-19 patients [66].

Brazilian studies can be examples of increased carbapenem resistance in *A. baumannii* isolates trends in Latin American territories in Pandemic times. A recent report described that CRAb was notified in 7.9% (373/4734) of device-associated infections notifications in 2019 and 12.4% (805/6514) in 2020 in 99 hospitals from Paraná state, south of Brazil. The monthly incidence density of CRAB per 1000 patient days increased significantly after April 2020, having a strong positive correlation with the incidence density of COVID-19 [67].

Polly et al. reported a retrospective observational study that compared the incidence density of HAIs caused by MDR bacteria (including CRAb) pre-COVID (2017–2019) and during the COVID-19 pandemic (2020) in hospitalized patients at a tertiary care public teaching hospital (São Paulo, Brazil). CRAb incidence density in the Pre-pandemic period (2017–2019) was 0.53. That increase can also be expressed by 108% in HAI infection by CRAB in all hospital units and 42% in ICU [68].

3. Mechanisms of carbapenem-resistance in *A. baumannii*

Several mechanisms of carbapenem resistance have been described in *A. baumannii* [1, 12, 48, 69]. Considering intrinsic cellular mechanisms, carbapenem resistance might be attributed to loss or decrease in outer membrane porins (OMPs), decreased drug affinity due to the downregulation of PBPs, and over-expression of efflux pumps [70–73].

However, inactivation or enzymatic degradation of carbapenems has been considered the major key associated with the development of carbapenem resistance in *A. baumannii* [12, 74, 75]. Different classes of carbapenem-hydrolyzing enzymes (carbapenemases) are based on molecular Ambler classification: Class A, B, and D [1, 12]. These enzymes are found frequently on plasmids and are transmissible [76]. Class A carbapenemases consist mainly of six members (SME, IMI, NMC, GES, SFC, and KPC), and GES class A carbapenemases seem the most prevalent in *A. baumannii*. Metallo-lactamases (MBLs), also called Class B enzymes, are potent carbapenemases, and four families (IMP, VIM, SIM, and NDM) have also been described in *A. baumannii* [77].

Instead of Class A and B, which are commonly identified in other bacterial pathogens, the carbapenem-hydrolyzing-class-D β-lactamases (CHDLs), also called oxacillinases/OXA-type β-lactamases, are referred as the most common carbapenemases in *A. baumannii* [12, 69]. β-lactamases of Ambler class D, OXA enzymes, possess an active serine site similar to class A and C β-lactamases. These β-lactamases show cloxacillin- and oxacillin-hydrolyzing activity and are classified into Bush-Jacoby functional group 2d. Those OXA enzymes that hydrolyze carbapenems belong to the Bush-Jacoby active subgroup 2df. Originally, OXA-type carbapenemases have mainly been found on the chromosomes of *A. baumannii* strains. However, several types of β-lactamases are also encoded on plasmids, allowing for their wide dissemination [78, 79].

There are six main groups in /OXA-type β-lactamases known to be harbored by *A. baumannii*: the intrinsic OXA-51-like and the acquired OXA-23-like, OXA-58-like, OXA-24/40-like, OXA-143-like and OXA-235-like [1, 12, 48, 80, 81]. Among them, OXA-23-like is the most prevalent worldwide. Clonal outbreaks of carbapenem-resistant and OXA–23–23-producing *A. baumannii* have been reported in many countries [82]. Analysis of the genetic environment of OXA-carbapenemases genes has shown that the genes are associated with various mobile elements [83].

A major expression of OXA genes might be facilitated by insertion sequences (ISs) because these genetic elements have strong promoters that enable the expression of OXA genes [74, 84]. For example, ISAba1, ISAba2, ISAba3, ISAba4, and IS18 are commonly associated with the presentation of carbapenemase genes in *A. baumannii* [85].

Countries	OXa-type carbapenemases	Other carbapenemases	References
Argentina	OXA-23-like, OXA-58-like	NDM	[87–92]
Bolivia	OXA-23-like, OXA-58-like	—	[93–96]
Brazil	OXA-23-like, OXA-24-like, OXA-58-like, OXA-143-like	KPC, NDM, IMP, VIM	[97–105]
Colombia	OXA-23-like, OXA-24-like, OXA-143-like	NDM, VIM	[52, 106–112]
Cuba	OXA-23-like, OXA-24-like, OXA-58-like	NDM	[113, 114]
Chile	OXA-23-like, OXA-58-like	—	[94, 115, 116]
Ecuador	OXA-23-like, OXA-24-like	NDM	[94, 117]
Honduras	—	NDM	[94, 118]
México	OXA-24-like, OXA-58-like, OXA-235-ilke	VIM	[80, 119–121]
Paraguay	OXA-23-like	—	[94]
Peru	OXA-23-like, OXA-24-like, OXA-143-like	NDM	[19, 122–125]
Puerto Rico	—	KPC	[97, 126, 127]
Uruguay	OXA-23-like, OXA-58-like	—	[94, 128]
Venezuela	OXA-23-like, OXA-58-like	NDM	[129–131]

Table 1.
Reports of the carbapenemase distribution in A. baumannii isolates in Latin America.

Transposons are another important genetic element responsible for the rapid spread of resistance genes worldwide [84]. The dissemination of bla_{OXA-23}, for example, has been strongly associated with transposons such as Tn2006, Tn2007, and Tn2008 that were identified as genetic structures harboring this gene [82, 85].

3.1 CRAb and carbapenemases in Latin America

More than 50% of *Acinetobacter* spp. Isolates in Latin America expressed carbapenem resistance. Additionally, the high prevalence of OXAs in CRAB isolates in Latin America is notorious [86]. Other carbapenemases have been reported in some Latin American countries but less frequently (**Table 1**). The spread of OXA-23 is also observed in Latin America and other parts of the world. And this dissemination has been commonly associated with CC113/CC79 and CC104/CC15 [132, 133].

4. Conclusion

In summary, this chapter presents a comprehensive review of the distribution of CRAb in Latin America. The chapter begins by highlighting the current significance of CRAb as a relevant pathogen associated with healthcare-acquired infections globally. Carbapenems have played a critical role as a therapeutic option for infections caused by MDR *A. baumannii*. However, the world has faced increased *A. baumannii* strains that have acquired carbapenem resistance. The spread of CRAb is associated with two international clones, IC 4 and IC 5 in the Latin countries. As observed in other parts of the world, carbapenem resistance is mediated mainly by OXA-type β-lactamases in Latin America. That dissemination illustrates these OXA-23-CRAb strains' success in Latin territory. Knowing the Latin American real scenario of CRAb is the first step in adopting measures to combat and control this challenging pathogen.

Acknowledgements

This research was funded by the Carlos Chagas Filho Foundation for Research Support of the State of Rio de Janeiro/FAPERJ (#110.198-13) and the Brazilian Council for Scientific Research (CNPq, #467.488/2014-2 and 301744/2019-0). Funding was also provided by FAPERJ (#210.003/2018) through the National Institutes of Science and Technology Program (INCT) to Carlos M. Morel (INCT-IDPN).

Author contributions

Conceived and designed the experiments: T.P.G.C., K.R.; writing—original draft: T.P.G.C.; review and editing: K.R., S.G.D.-S.; funding: S.G.D.-S. All authors have read and agreed to the published version of the manuscript.

Conflict of interest

The authors declare no conflict of interest.

Author details

Thiago Pavoni Gomes Chagas[1,2*], Karyne Rangel[3*] and
Salvatore Giovanni De-Simone[3,4*]

1 Medical School, Department of Pathology, Universidade Federal Fluminense (UFF), Niterói, RJ, Brazil

2 Molecular Epidemiology and Biotechnology Laboratory, Universidade Federal Fluminense (UFF), Niterói, RJ, Brazil

3 FIOCRUZ, Center for Technological Development in Health (CDTS)/National Institute of Science and Technology for Innovation in Neglected Population Diseases (INCT-IDPN), Rio de Janeiro, RJ. Brazil

4 Department of Molecular and Cellular Biology, Biology Institute, Universidade Federal Fluminense (UFF), Niterói, RJ, Brazil

*Address all correspondence to: tpgchagas@id.uff.br, karyne.rangelk@gmail.com and salvatore.simone@fiocruz.br

References

[1] Peleg AY, Seifert H, Paterson DL. *Acinetobacter baumannii*: Emergence of a successful pathogen. Clinical Microbiology Reviews. 2008;**21**(3):538-582. DOI: 10.1128/CMR.00058-07

[2] Ayoub Moubareck C, Hammoudi HD. Insights into *Acinetobacter baumannii*: A review of microbiological, virulence, and resistance traits in a threatening nosocomial pathogen. Antibiotics (Basel). 2020;**9**(3):119. DOI: 10.3390/antibiotics9030119

[3] Ibrahim S, Al-Saryi N, Al-Kadmy IMS, Aziz SN. Multidrug-resistant *Acinetobacter baumannii* as an emerging concern in hospitals. Molecular Biology Reports. 2021;**48**(10):6987-6998. DOI: 10.1007/s11033-021-06690-6

[4] Krishnaraj A, Gupta P, Bhatia M, Omar BJ. Role of matrix-assisted laser desorption ionization-time-of-flight mass spectrometry for species identification of *Acinetobacter* strains. Journal of Laboratory Physicians. 2023;**15**(3):336-343. DOI: 10.1055/s-0042-1760401

[5] Cisneros JM, Rodríguez-Baño J. Nosocomial bacteremia due to *Acinetobacter baumannii*: Epidemiology, clinical features and treatment. Clinical Microbiology and Infection. 2002;**8**(11):687-693. DOI: 10.1046/j.1469-0691.2002.00487.x

[6] Antunes LC, Visca P, Towner KJ. *Acinetobacter baumannii*: Evolution of a global pathogen. Pathogens and Disease. 2014;**71**(3):292-301. DOI: 10.1111/2049-632X.12125

[7] Zarrilli R, Pournaras S, Giannouli M, Tsakris A. Global evolution of multidrug-resistant *Acinetobacter baumannii* clonal lineages. International Journal of Antimicrobial Agents. 2013;**41**(1):11-19. DOI: 10.1016/j.ijantimicag.2012.09.008

[8] Fishbain J, Peleg AY. Treatment of *Acinetobacter* infections. Clinical Infectious Diseases. 2010;**51**(1):79-84. DOI: 10.1086/653120

[9] Zarrilli R, Bagattini M, Migliaccio A, Esposito EP, Triassi M. Molecular epidemiology of carbapenem-resistant *Acinetobacter baumannii* in Italy. Annali di Igiene. 2021;**33**(5):401-409. DOI: 10.7416/ai.2020.2395

[10] Durante-Mangoni E, Utili R, Zarrilli R. Combination therapy in severe *Acinetobacter baumannii* infections: An update on the evidence to date. Future Microbiology. 2014;**9**(6):773-789. DOI: 10.2217/fmb.14.34

[11] Yadav SK, Bhujel R, Hamal P, Mishra SK, Sharma S, Sherchand JB. Burden of multidrug-resistant *Acinetobacter baumannii* infection in hospitalized patients in a tertiary care hospital of Nepal. Infection and Drug Resistance. 2020;**13**:725-732. DOI: 10.2147/IDR.S239514

[12] Poirel L, Nordmann P. Carbapenem resistance in *Acinetobacter baumannii*: Mechanisms and epidemiology. Clinical Microbiology and Infection. 2006;**12**(9):826-836. DOI: 10.1111/j.1469-0691.2006.01456.x

[13] Spellberg B, Rex JH. The value of single-pathogen antibacterial agents. Nature Reviews. Drug Discovery. 2013;**12**(12):963. DOI: 10.1038/nrd3957-c1

[14] Piperaki ET, Tzouvelekis LS, Miriagou V, Daikos GL.

Carbapenem-resistant *Acinetobacter baumannii*: in pursuit of an effective treatment. Clinical Microbiology and Infection. 2019;**25**(8):951-957. DOI: 10.1016/j.cmi.2019.03.014

[15] Labarca JA, Salles MJ, Seas C, Guzmán-Blanco M. Carbapenem resistance in *Pseudomonas aeruginosa* and *Acinetobacter baumannii* in the nosocomial setting in Latin America. Critical Reviews in Microbiology. 2016;**42**(2):276-292. DOI: 10.3109/1040841X.2014.940494

[16] Antimicrobial Resistance Collaborators. Global burden of bacterial antimicrobial resistance in 2019: A systematic analysis. Lancet. 2022;**399**:629-655. DOI: 10.1016/S0140-6736(21)02724-0

[17] de Kraker ME, Stewardson AJ, Harbarth S. Will 10 million people die a year due to antimicrobial resistance by 2050? PLoS Medicine. 2016;**13**(11):e1002184. DOI: 10.1371/journal.pmed.1002184

[18] National Office for Animal Health. NOAH Response to Final O'Neill AMR Review Report July 2016. Middlesex: National Office for Animal Health; 2016 Available from: https://www.noah.co.uk/wp-content/uploads/2016/07/FINAL-NOAH-response-to-final-O-Neill-review-25-07-16-cle.pdf [Accessed:August 9, 2023]

[19] Levy-Blitchtein S, Roca I, Plasencia-Rebata S, Vicente-Taboada W, Velásquez-Pomar J, Muñoz L, et al. Emergence and spread of carbapenem-resistant *Acinetobacter baumannii* international clones II and III in Lima, Peru. Emerging Microbes & Infections. 2018;**7**(1):119. DOI: 10.1038/s41426-018-0127-9

[20] Pendleton JN, Gorman SP, Gilmore BF. Clinical relevance of the ESKAPE pathogens. Expert Review of Anti-Infective Therapy. 2013;**11**(3):297-308. DOI: 10.1586/eri.13.12

[21] Rello J, Kalwaje Eshwara V, Lagunes L, Alves J, Wunderink RG, Conway-Morris A, et al. A global priority list of the TOp TEn resistant microorganisms (TOTEM) study at intensive care: A prioritization exercise based on multi-criteria decision analysis. European Journal of Clinical Microbiology & Infectious Diseases. 2019;**38**(2):319-323. DOI: 10.1007/s10096-018-3428-y

[22] Tacconelli E, Carrara E, Savoldi A, Harbarth S, Mendelson M, Monnet DL, et al. Discovery, research, and development of new antibiotics: The WHO priority list of antibiotic-resistant bacteria and tuberculosis. The Lancet Infectious Diseases. 2018;**18**(3):318-327. DOI: 10.1016/S1473-3099(17)30753-3

[23] Mea HJ, Yong PVC, Wong EH. An overview of *Acinetobacter baumannii* pathogenesis: Motility, adherence and biofilm formation. Microbiological Research. 2021;**247**:126722. DOI: 10.1016/j.micres.2021.126722

[24] World Health Organization. WHO Publishes List of Bacteria for Which New Antibiotics Are Urgently Needed Internet. Geneva: WHO; 2017. Available from: https://www.who.int/news-room/detail/27-02-2017-who-publishes-list-of-bacteria-for-which-newantibiotics-are-urgently-needed [Accessed: September 9, 2023]

[25] Doi Y, Murray GL, Peleg AY. *Acinetobacter baumannii*: Evolution of antimicrobial resistance-treatment options. Seminars in Respiratory and Critical Care Medicine. 2015;**36**(1):85-98. DOI: 10.1055/s-0034-1398388

[26] Lee CR, Lee JH, Park M, Park KS, Bae IK, Kim YB, et al. Biology of *Acinetobacter baumannii*: Pathogenesis,

antibiotic resistance mechanisms, and prospective treatment options. Frontiers in Cellular and Infection Microbiology. 2017;7:55. DOI: 10.3389/fcimb.2017.00055

[27] Nguyen M, Joshi SG. Carbapenem resistance in *Acinetobacter baumannii*, and their importance in hospital-acquired infections: A scientific review. Journal of Applied Microbiology. 2021;**131**(6):2715-2738. DOI: 10.1111/jam.15130

[28] Papp-Wallace KM, Endimiani A, Taracila MA, Bonomo RA. Carbapenems: past, present, and future. Antimicrobial Agents and Chemotherapy. 2011;**55**(11):4943-4960. DOI: 10.1128/AAC.00296-11

[29] Bush K, Bradford PA. β-Lactams and β-lactamase inhibitors: An overview. Cold Spring Harbor Perspectives in Medicine. 2016;**6**(8):a025247. DOI: 10.1101/cshperspect.a025247

[30] Breilh D, Texier-Maugein J, Allaouchiche B, Saux MC, Boselli E. Carbapenems. Journal of Chemotherapy. 2013;**25**(1):1-17. DOI: 10.1179/1973947812Y.0000000032

[31] Bassetti M, Nicolini L, Esposito S, Righi E, Viscoli C. Current status of newer carbapenems. Current Medicinal Chemistry. 2009;**16**(5):564-575. DOI: 10.2174/092986709787458498

[32] Armstrong T, Fenn SJ, Hardie KR. JMM profile: Carbapenems: A broad-spectrum antibiotic. Journal of Medical Microbiology. 2021;**70**(12):001462. DOI: 10.1099/jmm.0.001462

[33] Shah PM. Parenteral carbapenems. Clinical Microbiology and Infection. 2008;**14**(Suppl 1):175-180. DOI: 10.1111/j.1469-0691.2007.01868.x. Erratum in: Clin Microbiol Infect. 2008;14(Suppl 5):21-24

[34] Kattan JN, Villegas MV, Quinn JP. New developments in carbapenems. Clinical Microbiology and Infection. 2008;**14**(12):1102-1111. DOI: 10.1111/j.1469-0691.2008.02101.x

[35] Yoon YK, Yang KS, Lee SE, Kim HJ, Sohn JW, Kim MJ. Effects of group 1 versus group 2 carbapenems on the susceptibility of *Acinetobacter baumannii* to carbapenems: A before and after intervention study of carbapenem-use stewardship. PLoS One. 2014;**9**(6):e99101. DOI: 10.1371/journal.pone.0099101

[36] Bradley JS, Garau J, Lode H, Rolston KV, Wilson SE, Quinn JP. Carbapenems in clinical practice: A guide to their use in serious infection. International Journal of Antimicrobial Agents. 1999;**11**(2):93-100. DOI: 10.1016/s0924-8579(98)00094-6

[37] Lima WG, Silva Alves GC, Sanches C, Antunes Fernandes SO, de Paiva MC. Carbapenem-resistant *Acinetobacter baumannii* in patients with burn injury: A systematic review and meta-analysis. Burns. 2019;**45**(7):1495-1508. DOI: 10.1016/j.burns.2019.07.006

[38] Bartal C, Rolston KVI, Nesher L. Carbapenem-resistant *Acinetobacter baumannii*: Colonization, infection and current treatment options. Infectious Disease and Therapy. 2022;**11**(2):683-694. DOI: 10.1007/s40121-022-00597-w

[39] Ma C, McClean S. Mapping global prevalence of *Acinetobacter baumannii* and recent vaccine development to tackle it. Vaccines (Basel). 2021;**9**(6):570. DOI: 10.3390/vaccines9060570

[40] European Centre for Disease Prevention and Control. WHO Regional Office for Europe/European Centre for Disease Prevention and Control. ECDPC. Antimicrobial Resistance Surveillance in

Europe 2022-2020 Data. 2022. Available from: https://www.ecdc.europa.eu/en/publications-data/antimicrobial-resistance-surveillance-europe-2022-2020-data [Accessed: July 10, 2023]

[41] Surveillance of Antimicrobial Resistance in Europe 2018: Annual Report of the European Antimicrobial Resistance Surveillance Network (EARS-Net) European Centre for Disease Prevention and Control; Solna, Sweden. 2019. Available from: https://www.ecdc.europa.eu/sites/default/files/documents/surveillance-antimicrobial-resistance-Europe-2018.pdf [Accessed: July 10, 2023]

[42] Centers for Disease Control and Prevention (CDC) Antibiotic Resistance Threats in the United States 2019 (Centers for Disease Control and Prevention) 150. 2019. Available from: https://www.cdc.gov/drugresistance/biggest-threats.html [Accessed: September 25, 2023]

[43] Gales AC, Seifert H, Gur D, Castanheira M, Jones RN, Sader HS. Antimicrobial susceptibility of *Acinetobacter calcoaceticus-Acinetobacter baumannii* Complex and *Stenotrophomonas maltophilia* clinical isolates: Results from the SENTRY antimicrobial surveillance program (1997-2016). Open Forum Infectious Diseases. 2019;**6**(Suppl 1):S34-S46. DOI: 10.1093/ofid/ofy293

[44] Seifert H, Blondeau J, Lucaßen K, Utt EA. Global update on the in vitro activity of tigecycline and comparators against isolates of *Acinetobacter baumannii* and rates of resistant phenotypes (2016-2018). Journal of Global Antimicrobial Resistance. 2022;**31**:82-89. DOI: 10.1016/j.jgar.2022.08.002

[45] Higgins PG, Dammhayn C, Hackel M, Seifert H. Global spread of carbapenem-resistant *Acinetobacter* baumannii. The Journal of Antimicrobial Chemotherapy. 2010;**65**(2):233-238. DOI: 10.1093/jac/dkp428. Epub 2009 Dec 8. Erratum in: J Antimicrob Chemother. 2010 Jun;65(6):1317

[46] Diancourt L, Passet V, Nemec A, Dijkshoorn L, Brisse S. The population structure of *Acinetobacter baumannii*: Expanding multiresistant clones from an ancestral susceptible genetic pool. PLoS One. 2010;**5**(4):e10034. DOI: 10.1371/journal.pone.0010034

[47] Al-Hassan L, Elbadawi H, Osman E, Ali S, Elhag K, Cantillon D, et al. Molecular epidemiology of carbapenem-resistant *Acinetobacter baumannii* from Khartoum state, Sudan. Frontiers in Microbiology. 2021;**12**:628736. DOI: 10.3389/fmicb.2021.628736

[48] Hamidian M, Nigro SJ. Emergence, molecular mechanisms and global spread of carbapenem-resistant *Acinetobacter baumannii*. Microbial Genomics. 2019;**5**(10):e000306. DOI: 10.1099/mgen.0.000306

[49] Nodari CS, Cayô R, Streling AP, Lei F, Wille J, Almeida MS, et al. Genomic analysis of carbapenem-resistant *Acinetobacter baumannii* isolates belonging to major endemic clones in South America. Frontiers in Microbiology. 2020;**11**:584603. DOI: 10.3389/fmicb.2020.584603

[50] Shelenkov A, Akimkin V, Mikhaylova Y. International clones of high risk of *Acinetobacter baumannii*-definitions, history, properties and perspectives. Microorganisms. 2023;**11**(8):2115. DOI: 10.3390/microorganisms11082115

[51] Castanheira M, Mendes RE, Gales AC. Global epidemiology and mechanisms of resistance of

Acinetobacter baumannii-calcoaceticus complex. Clinical Infectious Diseases. 2023;**76**(suppl 2):s166-S178. DOI: 10.1093/cid/ciad109

[52] Saavedra SY, Nuñez JC, Pulido IY, Gonzalez EB, Valenzuela EM, Reguero MT, et al. Characterisation of carbapenem-resistant *Acinetobacter calcoaceticus-a. baumannii* complex isolates in a third-level hospital in Bogotá, Colombia. International Journal of Antimicrobial Agents. 2008;**31**(4):389-391. DOI: 10.1016/j.ijantimicag.2007.12.008

[53] Chagas TP, Carvalho KR, de Oliveira Santos IC, Carvalho-Assef AP, Asensi MD. Characterization of carbapenem-resistant *Acinetobacter baumannii* in Brazil (2008-2011): Countrywide spread of OXA-23-producing clones (CC15 and CC79). Diagnostic Microbiology and Infectious Disease. 2014;**79**(4):468-472. DOI: 10.1016/j.diagmicrobio.2014.03.006

[54] Caldart RV, Fonseca EL, Freitas F, Rocha L, Vicente AC. *Acinetobacter baumannii* infections in Amazon region driven by extensively drug resistant international clones, 2016-2018. Memórias do Instituto Oswaldo Cruz. 2019;**114**:e190232. DOI: 10.1590/0074-02760190232

[55] Fonseca ÉL, Morgado SM, Freitas F, Oliveira PPC, Monteiro PM, Lima LS, et al. Persistence of a carbapenem-resistant *Acinetobacter baumannii* (CRAB) international clone II (ST2/IC2) sub-lineage involved with outbreaks in two Brazilian clinical settings. Journal of Infection and Public Health. 2023;**16**(10):1690-1695. DOI: 10.1016/j.jiph.2023.08.014

[56] Álvarez-Moreno C, Valderrama-Beltrán S, Rodriguez-Morales AJ. Implications of antibiotic use during the COVID-19 pandemic: The example of associated antimicrobial resistance in Latin America. Antibiotics (Basel). 2021;**10**(3):328. DOI: 10.3390/antibiotics10030328

[57] Rangel K, Chagas TPG, De-Simone SG. *Acinetobacter baumannii* infections in times of covid-19 pandemic. Pathogens. 2021;**10**(8):1006. DOI: 10.3390/pathogens10081006

[58] Shinohara DR, Dos Santos Saalfeld SM, Martinez HV, Altafini DD, Costa BB, Fedrigo NH, et al. Outbreak of endemic carbapenem-resistant *Acinetobacter baumannii* in a coronavirus disease 2019 (COVID-19)-specific intensive care unit. Infection Control and Hospital Epidemiology. 2022;**43**(6):815-817. DOI: 10.1017/ice.2021.98

[59] Sharifipour E, Shams S, Esmkhani M, Khodadadi J, Fotouhi-Ardakani R, Koohpaei A, et al. Evaluation of bacterial co-infections of the respiratory tract in COVID-19 patients admitted to ICU. BMC Infectious Diseases. 2020;**20**(1):646. DOI: 10.1186/s12879-020-05374-z

[60] Thomas GR, Corso A, Pasterán F, Shal J, Sosa A, Pillonetto M, et al. Increased detection of carbapenemase-producing Enterobacterales bacteria in Latin America and the Caribbean during the COVID-19 pandemic. Emerging Infectious Diseases. 2022;**28**(11):1-8. DOI: 10.3201/eid2811.220415

[61] Boral J, Genç Z, Pınarlık F, Ekinci G, Kuskucu MA, İrkören P, et al. The association between *Acinetobacter baumannii* infections and the COVID-19 pandemic in an intensive care unit. Scientific Reports. 2022;**12**(1):20808. DOI: 10.1038/s41598-022-25493-8

[62] Reyes LF, Rodriguez A, Fuentes YV, Duque S, García-Gallo E, Bastidas A, et al. Risk factors for developing

ventilator-associated lower respiratory tract infection in patients with severe COVID-19: A multinational, multicentre study, prospective, observational study. Scientific Reports. 2023;**13**(1):6553. DOI: 10.1038/s41598-023-32265-5

[63] Camargo CH, Yamada AY, Nagamori FO, de Souza AR, Tiba-Casas MR, de Moraes França FA, et al. Clonal spread of ArmA- and OXA-23-coproducing *Acinetobacter baumannii* international clone 2 in Brazil during the first wave of the COVID-19 pandemic. Journal of Medical Microbiology. 2022;**71**(4). DOI: 10.1099/jmm.0.001509. Available from: https://pubmed.ncbi.nlm.nih.gov/35417321/

[64] CastroMG UL, Vicino M, Cuevas G, Argarañá F. Rising incidence of carbapenem resistant isolates: An Argentinian hospital's experience. More trouble in the aftermath of the COVID-19 pandemic. Iberoamerican Journal of Medicine. 2022;**4**(2):92-99. DOI: 10.53986/ibjm.2022.0020

[65] Loyola-Cruz MÁ, Durán-Manuel EM, Cruz-Cruz C, Márquez-Valdelamar LM, Bravata-Alcántara JC, Cortés-Ortíz IA, et al. ESKAPE bacteria characterization reveals the presence of *Acinetobacter baumannii* and *Pseudomonas aeruginosa* outbreaks in COVID-19/VAP patients. American Journal of Infection Control. 2023;**51**(7):729-737. DOI: 10.1016/j.ajic.2022.08.012

[66] Alcántar-Curiel MD, Huerta-Cedeño M, Jarillo-Quijada MD, Gayosso-VázquezC,Fernández-VázquezJL, Hernández-Medel ML, et al. Gram-negative ESKAPE bacteria bloodstream infections in patients during the COVID-19 pandemic. PeerJ. 2023;**11**:e15007. DOI: 10.7717/peerj.15007

[67] de Carvalho Hessel Dias VM, Tuon F, de Jesus Capelo P, Telles JP, Fortaleza CMCB, Pellegrino Baena C. Trend analysis of carbapenem-resistant gram-negative bacteria and antimicrobial consumption in the post-COVID-19 era: An extra challenge for healthcare institutions. The Journal of Hospital Infection. 2022;**120**:43-47. DOI: 10.1016/j.jhin.2021.11.011

[68] Polly M, de Almeida BL, Lennon RP, Cortês MF, Costa SF, Guimarães T. Impact of the COVID-19 pandemic on the incidence of multidrug-resistant bacterial infections in an acute care hospital in Brazil. American Journal of Infection Control. 2022;**50**(1):32-38. DOI: 10.1016/j.ajic.2021.09.018

[69] Wong MH, Chan BK, Chan EW, Chen S. Over-expression of IS*Aba1*-linked intrinsic and exogenously acquired oxa type carbapenem-hydrolyzing-class D-ß-lactamase-encoding genes is key mechanism underlying carbapenem resistance in *Acinetobacter baumannii*. Frontiers in Microbiology. 2019;**10**:2809. DOI: 10.3389/fmicb.2019.02809

[70] Magnet S, Courvalin P, Lambert T. Resistance-nodulation-cell division-type efflux pump involved in aminoglycoside resistance in *Acinetobacter baumannii* strain BM4454. Antimicrobial Agents and Chemotherapy. 2001;**45**(12):3375-3380. DOI: 10.1128/AAC.45.12.3375-3380.2001

[71] Gribun A, Nitzan Y, Pechatnikov I, Hershkovits G, Katcoff DJ. Molecular and structural characterization of the HMP-AB gene encoding a pore-forming protein from a clinical isolate of *Acinetobacter baumannii*. Current Microbiology. 2003;**47**(5):434-443. DOI: 10.1007/s00284-003-4050-4

[72] Almasaudi SB. *Acinetobacter* spp. as nosocomial pathogens: Epidemiology and resistance features. Saudi. Journal of

Biological Sciences. 2018;**25**(3):586-596. DOI: 10.1016/j.sjbs.2016.02.009

[73] Siroy A, Molle V, Lemaître-Guillier C, Vallenet D, Pestel-Caron M, Cozzone AJ, et al. Channel formation by CarO, the carbapenem resistance-associated outer membrane protein of *Acinetobacter baumannii*. Antimicrobial Agents and Chemotherapy. 2005;**49**(12):4876-4883. DOI: 10.1128/AAC.49.12.4876-4883.2005

[74] Simo Tchuinte PL, Rabenandrasana MAN, Kowalewicz C, Andrianoelina VH, Rakotondrasoa A, Andrianirina ZZ, et al. Phenotypic and molecular characterisations of carbapenem-resistant *Acinetobacter baumannii* strains isolated in Madagascar. Antimicrobial Resistance and Infection Control. 2019;**8**:31. DOI: 10.1186/s13756-019-0491-9

[75] Abbott I, Cerqueira GM, Bhuiyan S, Peleg AY. Carbapenem resistance in *Acinetobacter baumannii*: Laboratory challenges, mechanistic insights and therapeutic strategies. Expert Review of Anti-Infective Therapy. 2013;**11**(4):395-409. DOI: 10.1586/eri.13.21

[76] Bansal G, Allen-McFarlane R, Eribo B. Antibiotic susceptibility, clonality, and molecular characterization of carbapenem-resistant clinical isolates of *Acinetobacter baumannii* from Washington DC. International Journal of Microbiology. 2020;**2020**:2120159. DOI: 10.1155/2020/2120159

[77] Nasiri MJ, Zamani S, Fardsanei F, Arshadi M, Bigverdi R, Hajikhani B, et al. Prevalence and mechanisms of carbapenem resistance in *Acinetobacter baumannii*: A comprehensive systematic review of cross-sectional studies from Iran. Microbial Drug Resistance. 2020;**26**(3):270-283. DOI: 10.1089/mdr.2018.0435

[78] Antunes NT, Fisher JF. Acquired class D β-lactamases. Antibiotics (Basel). 2014;**3**(3):398-434. DOI: 10.3390/antibiotics3030398

[79] Sawa T, Kooguchi K, Moriyama K. Molecular diversity of extended-spectrum β-lactamases and carbapenemases, and antimicrobial resistance. Journal of Intensive Care. 2020;**8**:13. DOI: 10.1186/s40560-020-0429-6

[80] Higgins PG, Pérez-Llarena FJ, Zander E, Fernández A, Bou G, Seifert H. OXA-235, a novel class D β-lactamase involved in resistance to carbapenems in *Acinetobacter baumannii*. Antimicrobial Agents and Chemotherapy. 2013;**57**(5):2121-2126. DOI: 10.1128/AAC.02413-12

[81] Abouelfetouh A, Torky AS, Aboulmagd E. Phenotypic and genotypic characterization of carbapenem-resistant *Acinetobacter baumannii* isolates from Egypt. Antimicrobial Resistance and Infection Control. 2019;**8**:185. DOI: 10.1186/s13756-019-0611-6

[82] Mugnier PD, Poirel L, Naas T, Nordmann P. Worldwide dissemination of the blaOXA-23 carbapenemase gene of *Acinetobacter baumannii*. Emerging Infectious Diseases. 2010;**16**(1):35-40. DOI: 10.3201/eid1601.090852

[83] Evans BA, Hamouda A, Amyes SGB. OXA-type β-lactamases in *Acinetobacter baumannii*: Emerging from the shadow of the extended-spectrum β-lactamases. Reviews and Research in Medical Microbiology. 2007;**18**(4):63-72. DOI: 10.1097/MRM.0b013e3282fce3d2

[84] Hashemizadeh Z, Hatam G, Fathi J, Aminazadeh F, Hosseini-Nave H, Hadadi M, et al. The spread of insertion sequences element and transposons in carbapenem resistant *Acinetobacter*

baumannii in a hospital setting in southwestern Iran. Infection & Chemotherapy. 2022;**54**(2):275-286. DOI: 10.3947/ic.2022.0022

[85] Pagano M, Martins AF, Barth AL. Mobile genetic elements related to carbapenem resistance in *Acinetobacter baumannii*. Brazilian Journal of Microbiology. 2016;**47**(4):785-792. DOI: 10.1016/j.bjm.2016.06.005

[86] Yu H, Ezpeleta-Lobato G, Han X, Carmona-Cartaya Y, Quiñones-Pérez D. Carbapenamase-producing *Acinetobacter baumannii* in China, Latin America and the Caribbean. MEDICC Review. 2022;**24**(1):59-69. DOI: 10.37757/MR2022.V24.N1.8

[87] Faccone D, Martino F, Pasteran F, Albornoz E, Biondi E, Vazquez M, et al. Multiple clones of metallo-β-lactamase-producing *Acinetobacter ursingii* in a children hospital from Argentina. Infection, Genetics and Evolution. 2019;**67**:145-149. DOI: 10.1016/j.meegid.2018.11.011

[88] Afzal-Shah M, Villar HE, Livermore DM. Biochemical characteristics of a carbapenemase from an *Acinetobacter baumannii* isolate collected in Buenos Aires, Argentina. Journal of Antimicrobial Chemotherapy. 1999;**43**(1):127-131. DOI: 10.1093/jac/43.1.127

[89] Merkier AK, Catalano M, Ramírez MS, Quiroga C, Orman B, Ratier L, et al. Polyclonal spread of *Bla*$_{(OXA\text{-}23)}$ and *Bla*$_{(OXA\text{-}58)}$ in *Acinetobacter baumannii* isolates from Argentina. Journal of Infection in Developing Countries. 2008;**2**(3):235-240. DOI: 10.3855/jidc.269

[90] Stietz MS, Ramírez MS, Vilacoba E, Merkier AK, Limansky AS, Centrón D, et al. *Acinetobacter baumannii* extensively drug resistant lineages in Buenos Aires hospitals differ from the international clones I-III. Infection, Genetics and Evolution. 2013;**14**:294-301. DOI: 10.1016/j.meegid.2012.12.020

[91] Rodríguez CH, Nastro M, Fiorilli G, Dabos L, Lopez Calvo J, Fariña ME, et al. Trends in the resistance profiles of *Acinetobacter baumannii* endemic clones in a university hospital of Argentina. Journal of Chemotherapy. 2016;**28**(1):25-27. DOI: 10.1179/1973947814Y.0000000213

[92] Traglia GM, Pasteran F, Escalante J, Nishimura B, Tuttobene MR, Subils T, et al. Genomic comparative analysis of two multi-drug resistance (MDR) *Acinetobacter baumannii* clinical strains assigned to international clonal lineage II recovered pre- and post-covid-19 pandemic. Biology (Basel). 2023;**12**(3):358. DOI: 10.3390/biology12030358

[93] Sennati S, Villagran AL, Bartoloni A, Rossolini GM, Pallecchi L. OXA-23-producing ST25 *Acinetobacter baumannii*: First report in Bolivia. Journal of Global Antimicrobial Resistance. 2016;**4**:70-71. DOI: 10.1016/j.jgar.2015.10.007

[94] Rodríguez CH, Balderrama Yarhui N, Nastro M, Nuñez Quezada T, Castro Cañarte G, Magne Ventura R, et al. Molecular epidemiology of carbapenem-resistant *Acinetobacter baumannii* in South America. Journal of Medical Microbiology. 2016;**65**(10):1088-1091. DOI: 10.1099/jmm.0.000328

[95] Sevillano E, Fernández E, Bustamante Z, Zabalaga S, Rosales I, Umaran A, et al. Emergence and clonal dissemination of carbapenem-hydrolysing OXA-58-producing *Acinetobacter baumannii* isolates in Bolivia. Journal of Medical Microbiology. 2012;**61**(Pt 1):80-84. DOI: 10.1099/jmm.0.032722-0

[96] Cerezales M, Xanthopoulou K, Wille J, Krut O, Seifert H, Gallego L, et al. Mobile genetic elements harboring antibiotic resistance determinants in *Acinetobacter baumannii* isolates from Bolivia. Frontiers in Microbiology. 2020;**11**:919. DOI: 10.3389/fmicb.2020.00919

[97] García-Betancur JC, Appel TM, Esparza G, Gales AC, Levy-Hara G, Cornistein W, et al. Update on the epidemiology of carbapenemases in Latin America and the Caribbean. Expert Review of Anti-Infective Therapy. 2021;**19**(2):197-213. DOI: 10.1080/14787210.2020.1813023

[98] Kiffer CRV, Rezende TFT, Costa-Nobre DT, Marinonio ASS, Shiguenaga LH, Kulek DNO, et al. A 7-year brazilian national perspective on plasmid-mediated carbapenem resistance in *Enterobacterales*, *Pseudomonas aeruginosa*, and *Acinetobacter baumannii* complex and the impact of the coronavirus disease 2019 pandemic on their occurrence. Clinical Infectious Diseases. 2023;**77**(Suppl 1):S29-S37. DOI: 10.1093/cid/ciad260

[99] Tognim MC, Gales AC, Penteado AP, Silbert S, Sader HS. Dissemination of IMP-1 metallo- beta -lactamase-producing *Acinetobacter* species in a Brazilian teaching hospital. Infection Control and Hospital Epidemiology. 2006;**27**(7):742-747. DOI: 10.1086/504356

[100] Cayô R, Rodrigues-Costa F, Matos AP, Carvalhaes CG, Jové T, Gales AC. Identification of a new integron harboring *Bla*$_{(IMP-10)}$ in carbapenem-resistant *Acinetobacter baumannii* clinical isolates. Antimicrobial Agents and Chemotherapy. 2015;**59**(6):3687-3689. DOI: 10.1128/AAC.04991-14

[101] Antonio CS, Neves PR, Medeiros M, Mamizuka EM, Elmor de Araújo MR, Lincopan N. High prevalence of carbapenem-resistant *Acinetobacter baumannii* carrying the blaOXA-143 gene in Brazilian hospitals. Antimicrobial Agents and Chemotherapy. 2011;**55**(3):1322-1323. DOI: 10.1128/AAC.01102-10

[102] Clímaco EC, Oliveira ML, Pitondo-Silva A, Oliveira MG, Medeiros M, Lincopan N, et al. Clonal complexes 104, 109 and 113 playing a major role in the dissemination of OXA-carbapenemase-producing *Acinetobacter baumannii* in Southeast Brazil. Infection, Genetics and Evolution. 2013;**19**:127-133. DOI: 10.1016/j.meegid.2013.06.024

[103] Vasconcelos AT, Barth AL, Zavascki AP, Gales AC, Levin AS, Lucarevschi BR, et al. The changing epidemiology of *Acinetobacter* spp. producing OXA carbapenemases causing bloodstream infections in Brazil: A BrasNet report. Diagnostic Microbiology and Infectious Disease. 2015;**83**(4):382-385. DOI: 10.1016/j.diagmicrobio.2015.08.006

[104] Pillonetto M, Arend L, Vespero EC, Pelisson M, Chagas TP, Carvalho-Assef AP, et al. First report of NDM-1-producing *Acinetobacter baumannii* sequence type 25 in Brazil. Antimicrobial Agents and Chemotherapy. 2014;**58**(12):7592-7594. DOI: 10.1128/AAC.03444-14

[105] Romanin P, Palermo RL, Cavalini JF, Fávaro LDS, De Paula-Petroli SB, Fernandes EV, et al. Multidrug- and extensively drug-resistant *Acinetobacter baumannii* in a tertiary hospital from Brazil: The importance of carbapenemase encoding genes and epidemic clonal complexes in a 10-year study. Microbial Drug Resistance. 2019;**25**(9):1365-1373. DOI: 10.1089/mdr.2019.0002

[106] Marquez-Ortiz RA, Haggerty L, Olarte N, Duarte C, Garza-Ramos U,

Silva-Sanchez J, et al. Genomic epidemiology of NDM-1-encoding plasmids in Latin American clinical isolates reveals insights into the evolution of multidrug resistance. Genome Biology and Evolution. 2017;**9**(6):1725-1741. DOI: 10.1093/gbe/evx115

[107] Gómez RF, Castillo A, Chávez-Vivas M. Characterization of multidrug-resistant *Acinetobacter* ssp. strains isolated from medical intensive care units in Cali–Colombia. Colombia Médica. 2017;**48**(4):183-190. DOI: 10.25100/cm.v48i4.2858

[108] Saavedra SY, Cayô R, Gales AC, Leal AL, Saavedra CH. Early dissemination of OXA-72-producing *Acinetobacter baumannii* strain in Colombia: A case report. The Brazilian Journal of Infectious Diseases. 2014;**18**(6):678-680. DOI: 10.1016/j.bjid.2014.05.017

[109] Vanegas JM, Higuita LF, Vargas CA, Cienfuegos AV, Rodríguez ÉA, Roncancio GE, et al. *Acinetobacter baumannii* resistente a carbapenémicos causante de osteomielitis e infecciones de la piel y los tejidos blandos *en hospitales de Medellín, Colombia [Carbapenem-resistant Acinetobacter* baumannii causing osteomyelitis and infections of skin and soft tissues in hospitals of Medellín, Colombia]. Biomédica. 2015;**35**(4):522-530. Spanish. DOI: 10.7705/biomedica.v35i4.2572

[110] Villegas MV, Kattan JN, Correa A, Lolans K, Guzman AM, Woodford N, et al. Dissemination of *Acinetobacter baumannii* clones with OXA-23 Carbapenemase in Colombian hospitals. Antimicrobial Agents and Chemotherapy. 2007;**51**(6):2001-2004. DOI: 10.1128/AAC.00226-07

[111] Saavedra SY, Prada-Cardozo D, Rincón V, Pérez-Cardona H, Hidalgo AM, González MN, et al. Whole-genome sequence of a colombian *Acinetobacter baumannii* strain, a coproducer of OXA-72 and OXA-255-like carbapenemases. Genome Announcements. 2017;**5**(7):e01558-e01516. DOI: 10.1128/genomeA.01558-16

[112] Correa A, Del Campo R, Escandón-Vargas K, Perenguez M, Rodríguez-Baños M, Hernández-Gómez C, et al. Distinct genetic diversity of carbapenem-resistant *Acinetobacter baumannii* from colombian hospitals. Microbial Drug Resistance. 2018;**24**(1):48-54. DOI: 10.1089/mdr.2016.0190

[113] Quiñones D, Carvajal I, Perez Y, Hart M, Perez J, Garcia S, et al. High prevalence of bla OXA-23 in *Acinetobacter* spp. and detection of bla_{NDM-1} in A. soli in Cuba: Report from National Surveillance Program (2010-2012). New Microbes New Infections. 2015;7:52-56. DOI: 10.1016/j.nmni.2015.06.002

[114] Quiñones Pérez D, Aung MS, Carmona Cartaya Y, González Molina MK, Pereda Novales N, Kobayashi N. Clonal diversity of *Acinetobacter* clinical isolates producing NDM-type carbapenemase in Cuba, 2013-19. IJID Regions. 2022;**5**:93-96. DOI: 10.1016/j.ijregi.2022.08.008

[115] Opazo-Capurro A, San Martín I, Quezada-Aguiluz M, Morales-León F, Domínguez-Yévenes M, Lima CA, et al. Evolutionary dynamics of carbapenem-resistant *Acinetobacter baumannii* circulating in Chilean hospitals. Infection, Genetics and Evolution. 2019;**73**:93-97. DOI: 10.1016/j.meegid.2019.04.022

[116] Opazo A, Domínguez M, Bello H, Amyes SG, González-Rocha G. OXA-type carbapenemases in *Acinetobacter baumannii* in South America. Journal

of Infection in Developing Countries. 2012;**6**(4):311-316. DOI: 10.3855/jidc.2310

[117] Villacís JE, Bovera M, Romero-Alvarez D, Cornejo F, Albán V, Trueba G, et al. NDM-1 carbapenemase in *Acinetobacter baumannii* sequence type 32 in Ecuador. New Microbes New Infections. 2019;**29**:100526. DOI: 10.1016/j.nmni.2019.100526

[118] López-Leal G, Zuniga-Moya JC, Castro-Jaimes S, Graña-Miraglia L, Pérez-Oseguera Á, Reyes-García HS, et al. Unexplored genetic diversity of multidrug- and extremely drug-resistant *Acinetobacter baumannii* isolates from tertiary hospitals in Honduras. Microbial Drug Resistance. 2019;**25**(5):690-695. DOI: 10.1089/mdr.2018.0311

[119] Alcántar-Curiel MD, Rosales-Reyes R, Jarillo-Quijada MD, Gayosso-Vázquez C, Fernández-Vázquez JL, Toledano-Tableros JE, et al. Carbapenem-resistant *Acinetobacter baumannii* in three tertiary care hospitals in Mexico: Virulence profiles, innate immune response and clonal dissemination. Frontiers in Microbiology. 2019;**10**:2116. DOI: 10.3389/fmicb.2019.02116

[120] Bocanegra-Ibarias P, Peña-López C, Camacho-Ortiz A, Llaca-Díaz J, Silva-Sánchez J, Barrios H, et al. *Genetic characterisation of drug resistance and clonal dynamics of Acinetobacter baumannii* in a hospital setting in Mexico. International Journal of Antimicrobial Agents. 2015;**45**(3):309-313. DOI: 10.1016/j.ijantimicag.2014.10.022

[121] Gonzalez-Villoria AM, Tamayo-Legorreta E, Garza-Ramos U, Barrios H, Sanchez-Pérez A, Rodríguez-Medina N, et al. A multicenter study in Mexico finds *Acinetobacter baumannii* clinical isolates belonging to clonal complexes 636B (113B) and 92B harboring OXA-72, OXA-239, and OXA-469. Antimicrobial Agents and Chemotherapy. 2016;**60**(4):2587-2588. DOI: 10.1128/AAC.02042-15

[122] Rocha C, Bernal M, Canal E, Rios P, Meza R, Lopez M, et al. First report of New Delhi Metallo-β-lactamase carbapenemase-producing *Acinetobacter baumannii* in Peru. The American Journal of Tropical Medicine and Hygiene. 2019;**100**(3):529-531. DOI: 10.4269/ajtmh.18-0802

[123] Yauri-Condor K, Zavaleta Apestegui M, Sevilla-Andrade CR, Villoslado Espinoza C, Taboada WV, Gonzales-Escalante E. Carbapenemasas clase D en aislamientos clínicos de *Acinetobacter baumannii*. Revista Peruana de Medicina Experimental y Salud Pública. 2020;**37**(2):387-388. Spanish. DOI: 10.17843/rpmesp.2020.372.4747

[124] Angles-Yanqui E, Huaringa-Marcelo J, Sacsaquispe-Contreras R, Pampa-Espinoza L. Panorama de las carbapenemasas en Perú Panorama of carbapenemases in PeruUm panorama das carbapenemases presentes no Peru. Revista Panamericana de Salud Pública. 2020;**44**:e61. DOI: https://doi.org/10.26633/RPSP.2020.61

[125] Castillo Y, Nieto C, Astocondor L, Jacobs J, Garcia C. Bacteriemia por *Acinetobacter baumannii* productor de oxacilinasa en hospitales de Lima, Perú [bacteremia caused by oxacillinase-producing *Acinetobacter baumannii* in hospitals in Lima, Peru]. Revista Peruana de Medicina Experimental y Salud Pública. 2019;**36**(2):364-366. Spanish. DOI: 10.17843/rpmesp.2019.362.4152

[126] Martinez T, Martinez I, Vazquez GJ, Aquino EE, Robledo IE. Genetic environment of the KPC gene in *Acinetobacter baumannii* ST2 clone from Puerto Rico and genomic insights into its drug resistance. Journal of Medical

Microbiology. 2016;**65**(8):784-792. DOI: 10.1099/jmm.0.000289

[127] Martínez T, Ropelewski AJ, González-Mendez R, Vázquez GJ, Robledo IE. Draft genome sequence of *Klebsiella pneumoniae* carbapenemase-producing *Acinetobacter baumannii* strain M3AC9-7, isolated from Puerto Rico. Genome Announcements. 2015;**3**(2):e00274-e00215. DOI: 10.1128/genomeA.00274-15

[128] Bado I, Papa-Ezdra R, Delgado-Blas JF, Gaudio M, Gutiérrez C, Cordeiro NF, et al. Molecular characterization of carbapenem-resistant *Acinetobacter baumannii* in the intensive care unit of Uruguay's university hospital identifies the first rmtC gene in the species. Microbial Drug Resistance. 2018;**24**(7):1012-1019. DOI: 10.1089/mdr.2017.0300

[129] Salazar De Vegas EZ, Nieves B, Ruiz M, Ruíz J, Vila J, María A, et al. Molecular epidemiology and characterization of resistance mechanisms to various antimicrobial agents in Acinetobacter baumannii isolated in Mérida, Venezuela. Medical Science Monitor. 2007;**13**(4):BR89-BR94

[130] Gomez-Gamboa L, Barrios-Camacho H, Duran-Bedolla J, Sánchez-Perez A, Reyna-Flores F, Perozo-Mena A, et al. Molecular and genetic characterization of carbapenemase-producing bacteria in Venezuela. Journal of Chemotherapy. 2019;**31**(6):349-353

[131] Gomez-Gamboa L, Perozo-Mena A, Bermudez-Gonzalez J, Villavicencio C, Villasmil J, Ginestre MM, et al. Detection of carbapenemase-producing bacteria in a public healthcare center from Venezuela. Journal of Infection in Developing Countries. 2020;**15**(1):163-167. DOI: 10.3855/jidc.13567

[132] Rodríguez CH, Nastro M, Famiglietti A. Carbapenemases in *Acinetobacter baumannii*. Review of their dissemination in Latin America. Revista Argentina de Microbiología. 2018;**50**(3):327-333. DOI: 10.1016/j.ram.2017.10.006

[133] Leal NC, Campos TL, Rezende AM, Docena C, Mendes-Marques CL, de Sá Cavalcanti FL, et al. Comparative genomics of *Acinetobacter baumannii* clinical strains from Brazil reveals polyclonal dissemination and selective exchange of mobile genetic elements associated with resistance genes. Frontiers in Microbiology. 2020;**11**:1176. DOI: 10.3389/fmicb.2020.01176